Analysis and Control of Output Synchronization for Complex Dynamical Networks

Jin-Liang Wang • Huai-Ning Wu • Tingwen Huang
Shun-Yan Ren

Analysis and Control of Output Synchronization for Complex Dynamical Networks

Springer

Jin-Liang Wang
School of Computer Science
and Software Engineering
Tianjin Polytechnic University
Tianjin, China

Tingwen Huang
Texas A&M University at Qatar
Doha, Qatar

Huai-Ning Wu
Beihang University
Beijing, China

Shun-Yan Ren
School of Mechanical Engineering
Tianjin Polytechnic University
Tianjin, China

ISBN 978-981-13-1351-6 ISBN 978-981-13-1352-3 (eBook)
https://doi.org/10.1007/978-981-13-1352-3

Library of Congress Control Number: 2018948210

This Springer imprint is published by the registered company Springer Nature Singapore Pte Ltd.
The registered company address is: 152 Beach Road, #21-01/04 Gateway East, Singapore 189721, Singapore

Preface

Recently, the synchronization problem for complex dynamical networks (CDNs) has stirred much research interest due to its fruitful applications in various fields. It is well known that the node state is difficult to be observed or measured, even the node state cannot be observed or measured at all. Moreover, sometimes only part states are needed to make synchronization to come true. For these phenomena, it is more interesting to study the output synchronization of CDNs. On the other hand, many real-world networks, such as transportation networks, social networks, communication networks, and so on, are coupled by multiple coupling forms. Obviously, it is also challenging and interesting to investigate the multi-weighted CDNs. Therefore, the investigation of output synchronization about CDNs with single weight and multiple weights has both practical and theoretical significance.

The aim of this book is to introduce recent research work on analysis and control of output synchronization for CDNs. The book is organized as follows:

Chapter 1: The background of single-weighted CDNs and multiple-weighted CDNs is introduced as well as the organization of this book, and some important definitions and useful lemmas are also provided in this chapter.

Chapter 2: This chapter investigates the output synchronization of a class of impulsive complex dynamical network with time-varying delay. By constructing suitable Lyapunov functionals, some new and useful conditions are obtained to guarantee the local and global exponential output synchronization of the impulsive CDNs. Finally, numerical examples are given to demonstrate the effectiveness of the theoretical results.

Chapter 3: This chapter considers a complex dynamical network model, in which the input and output vectors have different dimensions. By constructing appropriate Lyapunov functional, some sufficient conditions ensuring the passivity, input strict passivity, and output strict passivity are derived for the complex dynamical network with fixed coupling strength. In addition, we also reveal the relationship between output strict passivity and output synchronization of the complex dynamical network with fixed coupling strength. By employing the relationship between output strict passivity and output synchronization, a sufficient condition for output synchronization of the complex dynamical network with fixed coupling strength is established. Then, we extend these results to the case when the coupling strength is adaptively adjusted. Finally, two examples with numerical simulations are provided to demonstrate the effectiveness of the proposed criteria.

Chapter 4: This chapter focuses on the problem of output synchronization in undirected and directed CDNs, respectively, by applying Barbalat's lemma. First, to ensure the output synchronization, several sufficient criteria are established for these network models based on some mathematical techniques, such as the Lyapunov functional method and matrix theory. Furthermore, some adaptive schemes to adjust the coupling weights among network nodes are developed to achieve the output synchronization. By applying the designed adaptive laws, several criteria for output synchronization are deduced for the network models. In addition, a design procedure of the adaptive law is shown. Finally, two simulation examples are used to show the effectiveness of the previous results.

Chapter 5: This chapter studies the output synchronization of coupled neural networks (CNNs) as well as the effects of external disturbances. By employing matrix theory and Barbalat's lemma, several output synchronization criteria are presented for CNNs with directed and undirected topologies, respectively. Moreover, in order to ensure the output synchronization of CNNs, two adaptive schemes to adjust coupling weights are designed. On the other hand, we, respectively, analyze $\mathcal{H}_\infty$ output synchronization of directed and undirected CNNs with external disturbances, and two adaptive strategies for updating the coupling weights are designed to guarantee the $\mathcal{H}_\infty$ output synchronization of CNNs. Finally, two examples of CNNs are also given to verify the proposed output synchronization criteria.

Chapter 6: This chapter proposes a new complex dynamical network model with output coupling. This model is totally different from some existing complex dynamical network models. By exploiting Lyapunov functional method and some inequality techniques, several criteria on local and global exponential output synchronization are derived for the proposed network model. Finally, numerical simulations are given to illustrate the effectiveness of the derived results.

Chapter 7: In this chapter, two kinds of adaptive output synchronization problems for a complex delayed dynamical network with output coupling are investigated, that is, the cases with positive definite output matrix and with semi-positive definite output matrix. For the former, by using adaptive control method, a sufficient condition is obtained to guarantee the output synchronization of the complex dynamical network. In addition, a pinning adaptive output synchronization criterion is also derived for such network model. Then, we extend these results to the case when the output matrix is semi-positive definite. Finally, two numerical examples are provided to illustrate the effectiveness of the proposed results.

Chapter 8: In this chapter, we introduce two CDNs with multi-weights, which have several different sorts of weights between two nodes. By means of Lyapunov functional method and pinning control technique, some sufficient conditions are derived to ensure the synchronization for proposed network models. Moreover, some adaptive strategies are given to acquire suitable coupling strengths and feedback gains. By exploiting these designed adaptive laws, several general criteria for network synchronization are established. Finally, two numerical examples are also provided to show the validity of the theoretical results.

Chapter 9: The output synchronization and $\mathcal{H}_\infty$ output synchronization problems for multi-weighted complex dynamical network are discussed in this chapter. First, we analyze the output synchronization of multi-weighted complex dynamical network by exploiting Lyapunov functional and Barbalat's lemma. In addition, some nodes-based and edges-based pinning control strategies are developed to ensure the output synchronization of multi-weighted complex dynamical network. Similarly, the $\mathcal{H}_\infty$ output synchronization problem of multi-weighted complex dynamical network is also discussed. Finally, two numerical examples are presented to verify the correctness of the obtained results.

Acknowledgements

This book was supported by the National Natural Science Foundation for Distinguished Young Scholars of China under Grant 61625302, the National Natural Science Foundation of China under Grants 61773285, 61473011, and 61721091, the Program for Innovative Research Team in University of Tianjin (No. TD13-5032), and the National Priorities Research Program (NPRP) from the Qatar National Research Fund (a member of Qatar Foundation) under Grant NPRP 9-166-1-031. I'd like to begin by acknowledging my postgraduates Wei-Zhong Chen, Zhen Qin, Xiao-Xiao Zhang, Shui-Han Qiu, Qing Wang, Dong-Yang Wang, Jie Hou, and Shan-Rong Lin who have unselfishly given their valuable time in arranging these raw materials into something I'm proud of.

Tianjin, China — Jin-Liang Wang
Beihang, China — Huai-Ning Wu
Doha, Qatar — Tingwen Huang
Tianjin, China — Shun-Yan Ren
May 2018

Contents

Symbols and Acronyms

$\mathbb{N}$	$1, 2, 3, \cdots$
$\mathbb{R}$	field of real numbers
$\mathbb{R}^+$	field of nonnegative real numbers
$\mathbb{R}^n$	n-dimensional real Euclidean space
$\mathbb{R}^{m\times n}$	space of all $m \times n$ real matrices
I_n	$n \times n$ real identity matrix
$P > 0$	symmetric positive definite
$P \geqslant 0$	symmetric positive semi-definite
$P < 0$	symmetric negative definite
$P \leqslant 0$	symmetric negative semi-definite
B^{T}	transpose of matrix B
B^{-1}	inverse of matrix B
$\otimes$	Kronecker product of two matrices
$\lambda_M(A)$	maximum eigenvalue of matrix A
$\lambda_m(A)$	minimum eigenvalue of matrix A
$\mathrm{diag}(\cdots)$	block-diagonal matrix
$\|\| \cdot \|\|$	Euclidean norm of a vector and its induced norm of a matrix
sup	supremum
inf	infimum

$PC[[-\tau, +\infty), \mathbb{R}]$	$\{\phi : [-\tau, +\infty) \to \mathbb{R}, \phi(t)$ is continuous everywhere except for the points $t_k, k \in \mathbb{N}$ at which $\phi(t_k^+) = \phi(t_k)$ and $\phi(t_k^-)$ exist$\}$
$C([-\tau, 0], \mathbb{R}^n)$	a Banach space of continuous functions mapping the interval $[-\tau, 0]$ into $\mathbb{R}^n$ with the norm $\|\phi\|_\tau = \sup_{-\tau \leqslant \theta \leqslant 0} \|\phi(\theta)\|$
Rank$(\cdot)$	the rank of a matrix
CDNs	complex dynamical networks
CNNs	coupled neural networks
CDDNs	complex delayed dynamical networks

Chapter 1
Introduction

1.1 Background

1.1.1 CDNs with Single Weight

In the real world, complex networks can be seen everywhere, and have been viewed as a fundamental tool in understanding dynamical behavior and the response of real systems such as food webs, communication networks, social networks, cellular networks, World Wide Web, metabolic systems, disease transmission networks, biological neural networks, CNNs, power grids, and many others [1–3]. The investigation of CDNs obviously plays a prominent role both in application and theory, and the topology and dynamical behavior of various complex networks have been extensively studied by researchers [4, 5]. Especially, as one of the most significant and interesting dynamical properties of the complex networks, synchronization has received much of the focus in recent years. So far, a great many important results on synchronization have been obtained for various complex networks such as time invariant, time-varying, and impulsive network models; see [6–21] and relevant references therein.

It should be noticed that the state synchronization of complex networks with state coupling was considered in these articles (see also the above mentioned references). Practically, there are two kinds of coupling forms in complex networks: state coupling and output coupling. As we know, many phenomena in nature can be modeled as complex networks with output coupling [22, 23]. Nevertheless, there are very few works on CDNs with output coupling [22–24]. To our knowledge, Jiang, Tang and Chen [22] first introduced a complex network model with output coupling, and established some conditions for synchronization of such network model based on the Lyapunov stability theory. In [23], Chen proposed a complex network model with output coupling and random sensor delay, and a sufficient synchronization condition was given to ensure that the proposed network model is exponentially mean-square stable.

J.-L. Wang et al., *Analysis and Control of Output Synchronization for Complex Dynamical Networks*, https://doi.org/10.1007/978-981-13-1352-3_1

One should note that the state synchronization was investigated in [22, 23]. It is well known that the node state in complex networks is difficult to be observed or measured, even the node state can not be observed or measured at all. Moreover, in many circumstances, only part states are needed to make the synchronization to come true. For these phenomena, it is more interesting to study the output synchronization of complex networks. Practically, the output synchronization of complex networks has attracted great attention of researchers in recent years, and a large number of criteria have been derived to guarantee output synchronization for different complex network models [24–38]. For instance, Wang and Wu [25] discussed the output synchronization of a class of impulsive CDNs with time-varying delay, and some useful conditions were obtained to guarantee the local and global exponential output synchronization of the impulsive complex networks by constructing suitable Lyapunov functionals. In [26], the authors took the output synchronization into consideration for an impulsive complex delayed dynamical network, and some criteria were acquired to ensure global exponential output synchronization of the above-mentioned network by means of impulsive delay differential inequalities. By exploiting the relationship between output synchronization and output strict passivity, Wang et al. [29] presented two output synchronization criteria for the complex networks with adaptive and fixed coupling strengths. To the best of our knowledge, very few authors have considered the output synchronization for complex networks with output coupling [24, 30]. In [30], the authors considered a kind of complex networks that not only has time-varying delay but also has output coupling, and discussed the output synchronization of such network model. Therefore, it is essential to further study the output synchronization of CDNs with time-varying delay and output coupling.

1.1.2 CDNs with Multiple Weights

However, most of the existing results on synchronization are all based on network models with a single weight. As is well known, many real-world networks can be described by complex dynamical network models with multi-weights, where nodes are coupled by multiple coupling forms, such as transportation networks, social networks, communication networks, and so on. For instance, individuals or organizations can get in touch with each other by e-mail, mobile phone, facebook, etc. Considering that every way of communication has different weights, social networks should be modeled by multi-weighted complex networks. Furthermore, in the public traffic roads networks, taking every bus line as the network node, the edge means that two different bus lines have the same bus stops. In this case, there obviously exist different coupling weights in public traffic roads networks, such as coefficient of bus line length, passenger flow density, departing frequency, and so on [39]. Consequently, it is of great importance to investigate the multi-weighted CDNs. However, very few researchers have studied the synchronization problem of CDNs with multi-weights [39–41], specially the output synchronization problem. In

[39], the authors investigated the global synchronization of multi-weighted complex networks by changing transfers coefficient, congestion degrees and passenger flow density between different bus lines. An et al. [40] derived a criterion for global synchronization of the public traffic roads networks with multi-weights based on the Lyapunov stability theory. In [41], the authors considered a multi-weighted complex network with time-varying delays and uncertain inner coupling matrices, and a synchronization criterion was established for such network model by exploiting the designed adaptive feedback controller.

1.2 Book Organization

In recent years, more and more attention has been paid to the synchronization problem for CDNs due to its fruitful applications in various fields. As is well known, the node state is difficult to be observed or measured, even the node state can not be observed or measured at all. Moreover, sometimes only part states are needed to make synchronization to come true. On the other hand, many real-world networks, such as transportation networks, social networks, communication networks, and so on, are coupled by multiple coupling forms. Thus, the main aim of this book is to introduce some recent results on analysis and control of output synchronization for CDNs with single weight and multiple weights, and this book can serve as a stepping stone to study the output synchronization of CDNs. The rest of this book is organized as follows:

Chapter 2 introduces a new impulsive complex dynamical network model with time-varying delay, and investigates the local and global exponential output synchronization of the proposed network. By constructing suitable Lyapunov functionals, some new and useful conditions are obtained to guarantee the local and global exponential output synchronization of the impulsive complex dynamical network.

Chapter 3 proposes a complex dynamical network with different dimensions of input and output. The passivity, input strict passivity and output strict passivity of the proposed network model with fixed and adaptive coupling strength are taken into consideration, and several sufficient conditions are established. Furthermore, the relationship between output strict passivity and output synchronization of the proposed network model with fixed and adaptive coupling strength are revealed. In addition, some sufficient conditions for ensuring the output synchronization are derived by utilizing the relationship between output strict passivity and output synchronization.

Chapter 4 focuses on the problem of the output synchronization in complex network models with directed and undirected topologies. By applying Barbalat's lemma and constructing the appropriate Lyapunov functionals, several output synchronization criteria are developed for these network models. Furthermore, some control schemes are presented to adjust the coupling weights among network nodes, and several criteria for the output synchronization are deduced by applying the designed adaptive laws.

Chapter 5 is concerned with the output synchronization and $\mathcal{H}_\infty$ output synchronization problems for directed and undirected CNNs. By employing matrix theory, Lyapunov functionals and Barbalat's Lemma, several output synchronization and $\mathcal{H}_\infty$ output synchronization criteria are presented for these network models. Moreover, some adaptive schemes to adjust coupling weights are developed for reaching output synchronization and $\mathcal{H}_\infty$ output synchronization.

Chapter 6 proposes a new complex delayed dynamical network model with output coupling, and studies the output synchronization of the proposed network model. Several local and global exponential output synchronization criteria are established for the complex delayed dynamical network model with output coupling.

Chapter 7 presents a new complex delayed dynamical network with output coupling and semi-positive definite output matrix. The adaptive output synchronization of the proposed network model is investigated, and some sufficient conditions are obtained to guarantee the output synchronization of the complex delayed dynamical network.

Chapter 8 discusses the pinning synchronization for directed and undirected CDNs with multi-weights, respectively. On one hand, some synchronization criteria are established for the multi-weighted complex network with directed topology, and several adaptive laws for tuning coupling strengths and feedback gains are presented. On the other hand, some similar results for undirected complex dynamical network with multi-weights also are derived.

Chapter 9 considers the multi-weighted complex dynamical network with and without external disturbances. With the help of the Lyapunov functionals and Barbalat's Lemma, we not only analyze the output synchronization of multi-weighted complex networks, but also present some nodes-based and edges-based pinning adaptive control strategies. On the other hand, a few similar results about the $\mathcal{H}_\infty$ output synchronization of multi-weighted complex networks with external disturbances are also obtained.

1.3 Some Definitions

Definition 1.1 (see [42]) A system with supply rate ϑ is said to be dissipative if there exists a nonnegative function $S : \mathbb{R}^+ \to \mathbb{R}^+$, called the storage function, such that

$$\int_{t_0}^{t_p} \vartheta(u, y)dt \geqslant S(t_p) - S(t_0)$$

for any $t_p, t_0 \in \mathbb{R}^+$ and $t_p \geqslant t_0$, where $u(t) \in \mathbb{R}^p$ and $y(t) \in \mathbb{R}^q$ are the input and output of the system at time t, respectively.

Definition 1.2 A system is said to be passive if it is dissipative with respect to

$$\vartheta(u, y) = y^T(t)Pu(t),$$

where $P \in \mathbb{R}^{q\times p}$ is a constant matrix, $u(t) \in \mathbb{R}^p$ and $y(t) \in \mathbb{R}^q$ are the input and output of the system at time t, respectively.

Definition 1.3 A system is said to be strictly passive if it is dissipative with respect to

$$\vartheta(u, y) = y^T(t)Pu(t) - y^T(t)Qy(t) - u^T(t)Wu(t)$$

for $W \in \mathbb{R}^{p\times p} \geqslant 0$, $Q \in \mathbb{R}^{q\times q} \geqslant 0$, $P \in \mathbb{R}^{q\times p}$ and $\lambda_m(W) + \lambda_m(Q) > 0$, where $u(t) \in \mathbb{R}^p$ and $y(t) \in \mathbb{R}^q$ are the input and output of the system at time t, respectively.

The system is said to be input-strictly passive if $W > 0$ and output-strictly passive if $Q > 0$.

Definition 1.4 (see [43]) Function class QUAD(Δ, P): let $P = \mathrm{diag}(p_1, p_2, \cdots, p_n)$ be a positive definite diagonal matrix and $\Delta = \mathrm{diag}(\delta_1, \delta_2, \cdots, \delta_n)$ be a diagonal matrix. QUAD(Δ, P) denotes a class of continuous functions $f(x,t) : \mathbb{R}^n \times [0, +\infty) \to \mathbb{R}^n$ satisfying

$$(x - y)^T P\{[f(x,t) - f(y,t)] - \Delta(x - y)\} \leqslant -\eta(x - y)^T(x - y)$$

for some $\eta > 0$, all $x, y \in \mathbb{R}^n$ and $t > 0$.

It can be verified that many of the benchmark chaotic systems belong to "Function class QUAD" [44], such as the Lorenz system [45], the Chen system [46] and the Lü system [47].

Definition 1.5 (see [1]) Let $A = (a_{ij})_{m\times n} \in \mathbb{R}^{m\times n}$ and $B = (b_{ij})_{p\times q} \in \mathbb{R}^{p\times q}$. Then the Kronecker product of A and B is defined as the matrix

$$A \otimes B = \begin{pmatrix} a_{11}B & a_{12}B & \cdots & a_{1n}B \\ a_{21}B & a_{22}B & \cdots & a_{2n}B \\ \vdots & \vdots & \cdots & \vdots \\ a_{m1}B & a_{m2}B & \cdots & a_{mn}B \end{pmatrix} \in \mathbb{R}^{mp\times nq}.$$

Definition 1.6 (see [48]) The matrix B of order n is said to be reducible if there is a permutation matrix $Q \in \mathbb{R}^{n\times n}$ satisfying

$$QBQ^T = \begin{pmatrix} B_1 & 0 \\ B_{21} & B_2 \end{pmatrix},$$

where B_1 and B_2 are square matrices of order at least one. If B is not reducible, then B is called irreducible.

1.4 Some Lemmas

Lemma 1.7 (Berman and Plemmons [49]) *Let* $A \in \mathbb{R}^{n\times n}$, *then*

$$\lambda_m(A)x^T x \leqslant x^T Ax \leqslant \lambda_M(A)x^T x$$

for any $x \in \mathbb{R}^n$ *if* A *is a symmetric matrix.*

Lemma 1.8 (see [50]) *If the differentiable function* $g(t)$ *has a finite limit as* $t \to +\infty$ *and if* $\dot{g}(t)$ *is uniformly continuous, then* $\dot{g}(t) \to 0$ *as* $t \to +\infty$.

Lemma 1.9 (see [51, 52]) *The following formulas show the important properties of the Kronecker product:*

$$(1)\ (M_1 \otimes M_2)^T = M_1^T \otimes M_2^T;$$

$$(2)\ (\gamma M_1) \otimes M_2 = M_1 \otimes (\gamma M_2);$$

$$(3)\ (M_1 + M_2) \otimes M_3 = M_1 \otimes M_3 + M_2 \otimes M_3;$$

$$(4)\ (M_1 \otimes M_2)(M_3 \otimes M_4) = (M_1 M_3) \otimes (M_2 M_4),$$

where $\gamma \in \mathbb{R}$, M_1, M_2, M_3, M_4 *are matrices with suitable dimensions.*

Lemma 1.10 (see [53]) *Suppose that* G *is irreducible; at same time, it satisfies* $\sum_{j=1}^{N} G_{ij} = 0$ *with* $G_{ij} \geqslant 0 (i \neq j)$. *Then, there exists a positive vector* $\theta = (\theta_1, \theta_2, \cdots, \theta_N)^T \in \mathbb{R}^N$ *such that*

$$(1)\ G^T\theta = 0;$$

$$(2)\ \hat{G} = \Xi G + G^T \Xi \text{ is symmetric and}$$

$$\sum_{j=1}^{N} \hat{G}_{ij} = \sum_{j=1}^{N} \hat{G}_{ji} = 0 \text{ for all } i = 1, 2, ..., N,$$

where $\Xi = \mathrm{diag}(\theta_1, \theta_2, \cdots, \theta_N)$.

Lemma 1.11 (see [54]) *Suppose that* $G = (G_{ij})_{N\times N}$ *is a real symmetric and irreducible matrix, where*

$$G_{ij} \geqslant 0 \ \ (i \neq j), \quad G_{ii} = -\sum_{\substack{j=1\\ j\neq i}}^{N} G_{ij}.$$

Then,

(1) *0 is an eigenvalue of matrix G with multiplicity 1 and all the other eigenvalues of G are strictly negative.*
(2) *The largest nonzero eigenvalue $\lambda_2(G)$ of the matrix G satisfies*

$$\lambda_2(G) = \max_{x^T \mathbf{1}_N = 0, x \neq 0} \frac{x^T G x}{x^T x}.$$

(3) *For any $\eta = (\eta_1, \eta_2, \cdots, \eta_N)^T \in \mathbb{R}^N$,*

$$\eta^T G \eta = -\frac{1}{2} \sum_{i=1}^{N} \sum_{j=1}^{N} G_{ij} (\eta_i - \eta_j)^2.$$

Lemma 1.12 (see [48]) *Let B be a matrix of order n. After that, there exists a permutation matrix Q of order n and an integer $m \geqslant 1$ satisfying*

$$QBQ^T = \begin{pmatrix} B_1 & B_{12} & \cdots & B_{1m} \\ 0 & B_2 & \cdots & B_{2m} \\ \vdots & \vdots & \ddots & \vdots \\ 0 & 0 & \cdots & B_m \end{pmatrix}, \tag{1.1}$$

where $B_1, B_2, \cdots, B_m$ are square irreducible matrices. The matrices $B_1, B_2, \cdots, B_m$ that occur as diagonal blocks in (1.1) *are uniquely determined to within simultaneous permutation of their lines, while their ordering in* (1.1) *is not necessarily unique.*

The form in (1.1) *is called the Frobenius normal form of the square matrix B.*

Lemma 1.13 (see [55]) *Assume that $G = (G_{ij})_{N \times N} (N \geqslant 2)$ is an irreducible matrix, where*

$$G_{ij} \geqslant 0 \ (i \neq j), \quad G_{ii} \leqslant -\sum_{\substack{j=1 \\ j \neq i}}^{N} G_{ij}.$$

After that, there exists a diagonal matrix $\mathcal{K} = \mathrm{diag}(\mathcal{K}_1, \mathcal{K}_2, \cdots, \mathcal{K}_N) \in \mathbb{R}^{N \times N} > 0$ such that

$$\mathcal{K}G + G^T \mathcal{K} \leqslant 0.$$

Chapter 2
Output Synchronization Criteria for Impulsive CDNs with Time-Varying Delay

2.1 Introduction

In the real world, complex networks can be seen everywhere, and have been viewed as a fundamental tool in understanding dynamical behavior and the response of real systems such as food webs, communication networks, social networks, power grids, cellular networks, World Wide Web, metabolic systems, disease transmission networks, and many others [1]. The topology and dynamical behavior of various complex networks have been extensively studied by researchers. In particular, the synchronization problem has received much of the focus in recent years. A wide variety of synchronization criteria have been presented for various complex networks, see [7, 9, 14, 19–21, 24, 38, 56–62] and relevant references therein.

It should be noticed that the node state in complex networks is difficult to be observed or measured, even the node state can not be observed or measured at all. Moreover, in many circumstances only part states are needed to make the synchronization to come true. For these phenomena, it is interesting to study the output synchronization of complex networks. To our knowledge, there are few works on the output synchronization of complex dynamical networks [24]. In [24], Wang and Wu proposed a new complex dynamical network model with output coupling, and studied the output synchronization of the proposed network model. By constructing suitable Lyapunov functionals, some sufficient conditions ensuring local and global exponential output synchronization were obtained.

On the other hand, impulsive phenomena can be found in a wide variety of evolutionary process. In particular, some biological systems such as biological neural networks and bursting rhythm models in pathology, as well as frequency-modulated signal processing systems, optimal control models in economics, and flying object motions, in which many sudden and sharp changes occur instantaneously, in the form of impulse. Besides, impulsive phenomena can also be found in other fields of information science, automatic control systems, electronics, robotics, computer networking, artificial intelligence, and telecommunications, etc. [63]. Recently,

J.-L. Wang et al., *Analysis and Control of Output Synchronization for Complex Dynamical Networks*, https://doi.org/10.1007/978-981-13-1352-3_2

some researchers have studied the state synchronization of impulsive complex dynamical networks [64–66]. Zhou et al. [64] investigated the state synchronization of complex dynamical networks with impulsive effects. They proved that state synchronization of complex networks is heavily dependent on impulsive effects of connecting configuration in the networks. In [65], Li and Lai studied adaptive-impulsive state synchronization of uncertain complex dynamical networks, and several synchronization criteria were established. To the best of our knowledge, the output synchronization of complex networks with impulsive effects and time-varying delay has not yet been studied.

Motivated by the above discussion, we propose an impulsive complex network model with time-varying delay. By constructing suitable Lyapunov functionals, some sufficient conditions on local and global exponential output synchronization are derived for the proposed network model.

The chapter is organized as follows. An impulsive complex network model is introduced in Sect. 2.2. Some local and global exponential output synchronization criteria are established in Sect. 2.3. In Sect. 2.4, numerical examples are provided to illustrate the effectiveness of the theoretical results. Finally, Sect. 2.5 concludes the investigation and proposes some further work.

2.2 Network Model

In this chapter, we consider an impulsive complex network consisting of N identical nodes with diffusive and delay coupling, in which each node is an n-dimensional dynamical system. The mathematical model of the impulsive complex network can be described as follows:

$$\begin{cases} \dot{x}_i(t) = f(x_i(t)) + \dfrac{a}{k_i^{\beta_\omega}} \displaystyle\sum_{j=1}^{N} L_{ij}\Gamma_1 x_j(t) \\ \qquad + \dfrac{b}{k_i^{\beta_\omega}} \displaystyle\sum_{j=1}^{N} L_{ij}\Gamma_2 x_j(t-\tau(t)), \quad t \neq t_k, \\ y_i(t) = C x_i(t), \\ \Delta x_i = I_k(x_i), \quad t = t_k, \end{cases} \tag{2.1}$$

where $i = 1, 2, \cdots, N$, $k \in \mathbb{N}$. $\tau(t)$ is the time-varying delay with $0 \leqslant \tau(t) \leqslant \tau$.

The function $f(\cdot)$, describing the local dynamics of the nodes, is continuously differentiable and capable of producing various rich dynamical behaviors, $x_i(t) = (x_{i1}(t), x_{i2}(t), \cdots, x_{in}(t))^T \in \mathbb{R}^n$ is the state variable of node i, $y_i(t) \in \mathbb{R}^n$ is the output of node i, C is known matrix with appropriate dimensions, Γ_1 and

$\Gamma_2 \in \mathbb{R}^{n\times n}$ describe the coupling relations between two nodes for non-delayed configuration and delayed one, respectively, a and b are positive real numbers, which represent the overall coupling strength for non-delayed configuration and delayed one, respectively, k_i is the degree of node i and β_ω is a tunable weight parameter, the real matrix $L = (L_{ij})_{N\times N}$ is a symmetric matrix with diagonal entries $L_{ii} = -k_i$ and off-diagonal entries $L_{ij} = 1$ if node i and node j are connected by a link, and $L_{ij} = 0$ otherwise. The fixed moments t_k satisfy $0 = t_0 < t_1 < t_2 < \cdots < t_k < \cdots$, and $\lim_{k\to+\infty} t_k = +\infty, k \in \mathbb{N}$. At time instants t_k, jumps in the state variable x_i are denoted by $\Delta x_i \mid_{t=t_k} = x_i(t_k) - x_i(t_k^-)$, where $x_i(t_k^-) = \lim_{t\to t_k^-} x_i(t)$, $I_k(x_i) \in \mathbb{R}^n$ denote the incremental change of the state variable x_i at time t_k.

For the impulsive complex network (2.1), its initial conditions are given by $x_i(s) = \Phi_i(s) \in C([-\tau, 0], \mathbb{R}^n), i = 1, 2, \cdots, N$. In this chapter, we always assume that C is nonsingular matrix and network (2.1) is connected.

Definition 2.1 The impulsive complex network (2.1) is said to achieve output synchronization if

$$\lim_{t\to+\infty} ||y_i(t) - y_j(t)|| = 0 \text{ for all } i, j = 1, 2, \cdots, N.$$

Remark 2.2 In this model, the parameter $\beta_\omega = 0$ recovers that the network is unweighted and undirected, and the condition $\beta_\omega \neq 0$ corresponds to a network with weighted configuration and bidirectional [58]. Moreover, we take into account the case that the coupling relations and the coupling strength are related to the current state and the delayed state.

2.3 Main Results

For the convenience, we denote

$$\hat{f}(x(t)) = Cf(C^{-1}x(t)),$$

where $x(t) \in \mathbb{R}^n$.

Throughout this chapter, it is assumed that $h(t) \in \mathbb{R}^n$ is an equilibrium point of the following system

$$\dot{x}(t) = \hat{f}(x(t)). \tag{2.2}$$

For convenient analysis, we let

$$G_{ij} = \frac{L_{ij}}{k_i^{\beta_\omega}},$$

then we can get

$$\begin{cases} \dot{y}_i(t) = \hat{f}(y_i(t)) + a \sum_{j=1}^{N} G_{ij} C \Gamma_1 C^{-1} y_j(t) \\ \qquad + b \sum_{j=1}^{N} G_{ij} C \Gamma_2 C^{-1} y_j(t - \tau(t)), & t \neq t_k, \\ \Delta y_i = I_k(y_i), & t = t_k, \\ y_i(s) = C \Phi_i(s), & s \in [-\tau, 0], \end{cases}$$

where $i = 1, 2, \cdots, N$, $k \in \mathbb{N}$, $I_k(y_i) = y_i(t_k) - y_i(t_k^-) = C I_k(x_i)$. Letting $e_i(t) = y_i(t) - h(t)$, we can obtain

$$\begin{cases} \dot{e}_i(t) = \hat{f}(y_i(t)) - \hat{f}(h(t)) + a \sum_{j=1}^{N} G_{ij} C \Gamma_1 C^{-1} e_j(t) \\ \qquad + b \sum_{j=1}^{N} G_{ij} C \Gamma_2 C^{-1} e_j(t - \tau(t)), & t \neq t_k, \\ \Delta e_i = I_k(e_i), & t = t_k, \\ e_i(s) = C \Phi_i(s) - h(s), & s \in [-\tau, 0], \end{cases} \tag{2.3}$$

where $k \in \mathbb{N}$, $I_k(e_i) = e_i(t_k) - e_i(t_k^-) = I_k(y_i)$. Its linearized system reads as

$$\begin{cases} \dot{e}_i(t) = D\hat{f}(h(t)) e_i(t) + a \sum_{j=1}^{N} G_{ij} C \Gamma_1 C^{-1} e_j(t) \\ \qquad + b \sum_{j=1}^{N} G_{ij} C \Gamma_2 C^{-1} e_j(t - \tau(t)), & t \neq t_k, \\ \Delta e_i = I_k(e_i), & t = t_k, \\ e_i(s) = C \Phi_i(s) - h(s), & s \in [-\tau, 0], \end{cases} \tag{2.4}$$

where $D\hat{f}(h(t)) = \hat{f}'(h(t)) \in \mathbb{R}^{n \times n}$ is the Jacobian of $\hat{f}(x(t))$ at $h(t)$. We can rewrite network (2.4) in a compact form as follows:

$$\begin{cases} \dot{e}(t) = [I_N \otimes Df(h(t))]e(t) + a[G \otimes (C\Gamma_1 C^{-1})]e(t) \\ \qquad + b[G \otimes (C\Gamma_2 C^{-1})]e(t - \tau(t)), & t \neq t_k, \\ \Delta e = I_k(e), & t = t_k, \\ e(s) = \hat{\Phi}(s) - H(s), & s \in [-\tau, 0], \end{cases} \tag{2.5}$$

where $G = (G_{ij})_{N \times N}$, and

$$\hat{\Phi}(s) = [(C\Phi_1(s))^T, (C\Phi_2(s))^T, \cdots, (C\Phi_N(s))^T]^T,$$

$$H(s) = (h^T(s), h^T(s), \cdots, h^T(s))^T,$$

$$e(t) = \left(e_1^T(t), e_2^T(t), \cdots, e_N^T(t)\right)^T,$$

$$\Delta e = [(\Delta e_1)^T, (\Delta e_2)^T, \cdots, (\Delta e_N)^T]^T,$$

$$I_k(e) = (I_k^T(e_1), I_k^T(e_2), \cdots, I_k^T(e_N))^T,$$

$$e(t - \tau(t)) = (e_1^T(t - \tau(t)), e_2^T(t - \tau(t)), \cdots, e_N^T(t - \tau(t)))^T.$$

Since network (2.1) is connected in the sense of having no isolated clusters, which means that the matrix K ($K = \text{diag}(k_1, k_2, \cdots, k_N)$ is the diagonal matrix of degree.) is non-singular, and the coupling matrix L is irreducible. It is obvious that we can obtain

$$G = K^{-\beta_\omega} L.$$

Using matrix identities, we know that the spectrum of asymmetric matrix G is equal to the spectrum of the symmetric matrix $W = K^{-\beta_\omega/2} L K^{-\beta_\omega/2}$, which is semi-negative definite. Then, we can easily obtain the eigenvalues of G are real-valued and are strictly negative except an eigenvalue 0 with multiplicity 1, and G can be diagonalizable. Therefore, there exists a nonsingular matrix $\phi = (\phi_1, \phi_2, \cdots, \phi_N) \in \mathbb{R}^{N \times N}$ such that

$$\phi^{-1} G \phi = \Lambda$$

with

$$\Lambda = \text{diag}(\lambda_1, \lambda_2, \cdots, \lambda_N).$$

$\lambda_i, i = 1, 2, \ldots, N$, are the eigenvalues of G and

$$0 = \lambda_1 > \lambda_2 \geqslant \lambda_3 \geqslant \cdots \geqslant \lambda_N.$$

Take a nonsingular transform

$$(\phi^{-1} \otimes I_n)e(t) = \nu(t) = \left(\nu_1^T(t), \nu_2^T(t), \cdots, \nu_N^T(t)\right)^T \in \mathbb{R}^{nN}.$$

It follows from (2.5) that

$$\begin{cases} \dot{\nu}(t) = [I_N \otimes D\hat{f}(h(t)) + a\Lambda \otimes (C\Gamma_1 C^{-1})]\nu(t) \\ \qquad + b[\Lambda \otimes (C\Gamma_2 C^{-1})]\nu(t - \tau(t)), & t \neq t_k, \\ \Delta\nu = I_k(\nu), & t = t_k, \\ \nu(s) = (\phi^{-1} \otimes I_n)[\hat{\Phi}(s) - H(s)], & s \in [-\tau, 0], \end{cases} \tag{2.6}$$

where

$$\begin{aligned} \nu(t - \tau(t)) &= (\nu_1^T(t - \tau(t)), \nu_2^T(t - \tau(t)), \cdots, \nu_N^T(t - \tau(t)))^T \in \mathbb{R}^{nN}, \\ I_k(\nu) &= (I_k^T(\nu_1), I_k^T(\nu_2), \cdots, I_k^T(\nu_N))^T \\ &= (\phi^{-1} \otimes I_n)[(CI_k(x_1))^T, \cdots, (CI_k(x_N))^T]^T, \\ \Delta\nu &= [(\Delta\nu_1)^T, (\Delta\nu_2)^T, \cdots, (\Delta\nu_N)^T]^T. \end{aligned}$$

In the following, two sufficient conditions are presented, ensuring the local exponential output synchronization of impulsive complex network (2.1).

Theorem 2.3 *Let $\dot{\tau}(t) \leqslant \sigma < 1$. If there exist matrices $P, Q > 0$ and constants $\varepsilon > 0, \mu > 0, M > 0, \eta_k \geqslant 0$, such that*

$$\begin{pmatrix} W & bP[\Lambda \otimes (C\Gamma_2 C^{-1})] \\ b[\Lambda \otimes (C\Gamma_2 C^{-1})]^T P & -(1-\sigma)e^{-\varepsilon\tau}Q \end{pmatrix} \leqslant 0, \tag{2.7}$$

$$\mu\tau \leqslant \inf\{t_k - t_{k-1}, k \in \mathbb{N}\}, \tag{2.8}$$

$$\|I_k(\nu_i)\| \leqslant \eta_k \|\nu_i(t_k^-)\|, \tag{2.9}$$

$$\max\{\beta_k, k \in \mathbb{N}\} \leqslant M < e^{\varepsilon\mu\tau}, \tag{2.10}$$

where $W = \varepsilon P + aP(\Lambda \otimes (C\Gamma_1 C^{-1})) + a(\Lambda \otimes (C\Gamma_1 C^{-1}))^T P + P(I_N \otimes D\hat{f}(h(t))) + (I_N \otimes D\hat{f}(h(t)))^T P + Q$, $\beta_k = 1 + \frac{(2\eta_k + \eta_k^2)\|P\|}{\lambda_m(P)}$, $k \in \mathbb{N}, i = 1, 2, \cdots, N$, then impulsive complex network (2.1) *achieves local exponential output synchronization.*

Proof Construct Lyapunov functional for system (2.6) as follows

$$V(t) = e^{\varepsilon t}\nu^T(t)P\nu(t) + \int_{t-\tau(t)}^{t} e^{\varepsilon s}\nu^T(s)Q\nu(s)ds. \tag{2.11}$$

According to the definition of $V(t)$ and Lemma 1.7, we can obtain

$$V(t) \geqslant e^{\varepsilon t}\lambda_m(P)\|\nu(t)\|^2. \tag{2.12}$$

The derivative of $V(t)$ along the solution of system (2.6) is given as follows:

$$\begin{aligned}
\dot{V}(t) \leqslant\ & \varepsilon e^{\varepsilon t}\nu^T(t)P\nu(t) + 2e^{\varepsilon t}\nu^T(t)P\{[I_N \otimes D\hat{f}(h(t)) + a\Lambda \otimes (C\Gamma_1 C^{-1})]\nu(t) \\
& + b[\Lambda \otimes (C\Gamma_2 C^{-1})]\nu(t-\tau(t))\} + e^{\varepsilon t}\nu^T(t)Q\nu(t) \\
& - (1-\sigma)e^{\varepsilon(t-\tau)}\nu^T(t-\tau(t))Q\nu(t-\tau(t)) \\
=\ & \begin{pmatrix} \nu(t) \\ \nu(t-\tau(t)) \end{pmatrix}^T \begin{pmatrix} W & bP[\Lambda \otimes (C\Gamma_2 C^{-1})] \\ b[\Lambda \otimes (C\Gamma_2 C^{-1})]^T P & -(1-\sigma)e^{-\varepsilon\tau}Q \end{pmatrix} \\
& \times \begin{pmatrix} \nu(t) \\ \nu(t-\tau(t)) \end{pmatrix} e^{\varepsilon t} \\
\leqslant\ & 0, \qquad t \neq t_k, \quad k \in \mathbb{N}.
\end{aligned} \tag{2.13}$$

Also

$$\begin{aligned}
V(t_k) =\ & e^{\varepsilon t_k^-}\left(\nu(t_k^-) + I_k(\nu)\right)^T P(\nu(t_k^-) + I_k(\nu)) + \int_{t_k^- - \tau(t_k^-)}^{t_k^-} e^{\varepsilon s}\nu^T(s)Q\nu(s)ds \\
\leqslant\ & V(t_k^-) + 2\eta_k e^{\varepsilon t_k^-}\|P\|\|\nu(t_k^-)\|^2 + \eta_k^2 e^{\varepsilon t_k^-}\|P\|\|\nu(t_k^-)\|^2 \\
\leqslant\ & \beta_k V(t_k^-).
\end{aligned} \tag{2.14}$$

By (2.13) and (2.14), we have

$$\begin{aligned}
V(t) \leqslant\ & V(0)\prod_{i=1}^{k-1}\beta_i \\
\leqslant\ & V(0)M^{k-1}, \quad t \in [t_{k-1}, t_k), \quad k \in \mathbb{N}.
\end{aligned} \tag{2.15}$$

Since $\mu\tau \leqslant \inf\{t_k - t_{k-1}, k \in \mathbb{N}\}$, one has $k - 1 \leqslant \frac{t_{k-1}}{\mu\tau}$, which implies

$$M^{k-1} \leqslant e^{\frac{lnM}{\mu\tau}t}, \quad t \in [t_{k-1}, t_k), k \in \mathbb{N}. \tag{2.16}$$

Then, we have

$$V(t) \leqslant V(0)e^{\frac{lnM}{\mu\tau}t}, \quad t \in [t_{k-1}, t_k), k \in \mathbb{N}. \tag{2.17}$$

On the other hand,

$$V(0) = \nu^T(0)P\nu(0) + \int_{-\tau(0)}^{0} e^{\varepsilon s}\nu^T(s)Q\nu(s)ds$$
$$\leqslant (\lambda_M(P) + \tau\lambda_M(Q))\|\Xi\|_\tau^2, \tag{2.18}$$

where $\Xi(s) = (\phi^{-1} \otimes I_n)[\hat{\Phi}(s) - H(s)]$. It follows from (2.12), (2.17) and (2.18) that

$$\|\nu(t)\| \leqslant \left(\frac{\lambda_M(P) + \tau\lambda_M(Q)}{\lambda_m(P)}\right)^{\frac{1}{2}} \|\Xi\|_\tau e^{-\frac{(\varepsilon - \frac{\ln M}{\mu\tau})t}{2}}.$$

According to (2.10), system (2.6) is globally exponentially stable about its zero solution. Obviously, we can conclude that impulsive complex network (2.1) achieves local exponential output synchronization. The proof is completed.

Theorem 2.4 *Let* $\dot{\tau}(t) \leqslant \sigma < 1$ *and* $\nu_i(t_k) = r_{ik}\nu_i(t_k^-)$.

(i) *If* $r_k^2 \leqslant 1$ *for all* k, *then impulsive complex network* (2.1) *achieves local exponential output synchronization if there exist matrices* $Q > 0$, $P = \text{diag}\{P_1, P_2, \ldots, P_N\}$, $P_i > 0$, *and a positive constant* ε, *such that*

$$\begin{pmatrix} W & bP[\Lambda \otimes (C\Gamma_2C^{-1})] \\ b[\Lambda \otimes (C\Gamma_2C^{-1})]^T P & -(1-\sigma)e^{-\varepsilon\tau}Q \end{pmatrix} \leqslant 0 \tag{2.19}$$

where $i = 1, 2, \cdots, N, k \in \mathbb{N}$, $W = \varepsilon P + aP(\Lambda \otimes (C\Gamma_1C^{-1})) + a(\Lambda \otimes (C\Gamma_1C^{-1}))^T P + P(I_N \otimes D\hat{f}(h(t))) + (I_N \otimes D\hat{f}(h(t)))^T P + Q, r_k^2 = \max\{r_{ik}^2,\ i = 1, 2, \cdots, N\}$.

(ii) *If there exist* k *such that* $r_k^2 > 1$, *then impulsive complex network* (2.1) *achieves local exponential output synchronization if LMI* (2.19) *and the following conditions hold*

$$\mu\tau \leqslant \inf\{t_k - t_{k-1}, k \in \mathbb{N}\},\ \ \mu > 0, \tag{2.20}$$
$$\max\{r_k^2, k \in \mathbb{N}\} \leqslant M < e^{\varepsilon\mu\tau}. \tag{2.21}$$

Proof Take the same Lyapunov functional $V(t)$ as in Theorem 2.3. By the proof of Theorem 2.3, we get that (2.12) and (2.13) hold. Furthermore,

$$V(t_k) = e^{\varepsilon t_k^-}\nu^T(t_k^-)R_kP\,R_k\nu(t_k^-) + \int_{t_k^- - \tau(t_k^-)}^{t_k^-} e^{\varepsilon s}\nu^T(s)Q\nu(s)ds$$
$$\leqslant r_k^2 e^{\varepsilon t_k^-}\nu^T(t_k^-)P\nu(t_k^-) + \int_{t_k^- - \tau(t_k^-)}^{t_k^-} e^{\varepsilon s}\nu^T(s)Q\nu(s)ds,$$

where $R_k = \mathrm{diag}(r_{1k}I_n, r_{2k}I_n, \cdots, r_{Nk}I_n)$. If $r_k^2 \leqslant 1$ for all k, we can get

$$V(t_k) \leqslant V(t_k^-).$$

Then, we have

$$V(t) \leqslant V(0). \tag{2.22}$$

From the definition of $V(0)$, there obviously exists a positive constant ϱ such that

$$V(0) \leqslant \varrho \|\Xi\|_\tau^2. \tag{2.23}$$

It follows from (2.12), (2.22) and (2.23) that

$$\|\nu(t)\| \leqslant \left(\frac{\varrho}{\lambda_m(P)}\right)^{\frac{1}{2}} \|\Xi\|_\tau e^{-\frac{\varepsilon t}{2}}.$$

Thus, we can conclude that the impulsive complex network (2.1) achieves local exponential output synchronization. On the other hand, if there exist k such that $r_k^2 > 1$, we can get

$$V(t_k) \leqslant r_k^2 V(t_k^-).$$

Then, we can obtain

$$\begin{aligned} V(t) &\leqslant V(0)\prod_{i=1}^{k-1} \vartheta_i^2 \\ &\leqslant V(0)M^{k-1}, \quad t \in [t_{k-1}, t_k), \quad k \in \mathbb{N}, \end{aligned}$$

where

$$\vartheta_k^2 = \begin{cases} r_k^2, & r_k^2 > 1, \quad k \in \mathbb{N}, \\ 1, & r_k^2 \leqslant 1, \quad k \in \mathbb{N}. \end{cases}$$

From (2.20), we have

$$V(t) \leqslant V(0)e^{\frac{\ln M}{\mu\tau}t}, \quad t \in [t_{k-1}, t_k), \quad k \in \mathbb{N}. \tag{2.24}$$

It follows from (2.12), (2.23) and (2.24) that

$$\|\nu(t)\| \leqslant \left(\frac{\varrho}{\lambda_m(P)}\right)^{\frac{1}{2}} \|\Xi\|_\tau e^{-\frac{(\varepsilon - \frac{\ln M}{\mu\tau})t}{2}}.$$

According to (2.21), system (2.6) is globally exponentially stable about its zero solution. Then, we can conclude that impulsive complex network (2.1) achieves local exponential output synchronization. The proof is completed.

Next we provide two global exponential output synchronization results for impulsive complex network (2.1).

Firstly, we can rewrite system (2.3) in a compact form as follows:

$$\begin{cases} \dot{e}(t) = \hat{F}(y(t)) - \hat{F}(h(t)) + aG \otimes (C\Gamma_1 C^{-1})e(t) \\ \qquad + bG \otimes (C\Gamma_2 C^{-1})e(t - \tau(t)), & t \neq t_k, \\ \Delta e = I_k(e), & t = t_k, \\ e(s) = \Xi_1(s), & s \in [-\tau, 0], \end{cases} \tag{2.25}$$

where

$$\hat{F}(y(t)) = (\hat{f}^T(y_1(t)), \hat{f}^T(y_2(t)), \cdots, \hat{f}^T(y_N(t)))^T,$$

$$\Xi_1(s) = \hat{\Phi}(s) - H(s),$$

$$\hat{F}(h(t)) = (\hat{f}^T(h(t)), \hat{f}^T(h(t)), \cdots, \hat{f}^T(h(t)))^T,$$

$$e(t) = \left(e_1^T(t), e_2^T(t), \cdots, e_N^T(t)\right)^T,$$

$$\Delta e = [(\Delta e_1)^T, (\Delta e_2)^T, \cdots, (\Delta e_N)^T]^T,$$

$$e(t - \tau(t)) = (e_1^T(t - \tau(t)), e_2^T(t - \tau(t)), \cdots, e_N^T(t - \tau(t)))^T,$$

$$I_k(e) = (I_k^T(e_1), I_k^T(e_2), \cdots, I_k^T(e_N))^T,$$

$$H(s) = (h^T(s), h^T(s), \cdots, h^T(s))^T,$$

$$\hat{\Phi}(s) = [(C\Phi_1(s))^T, (C\Phi_2(s))^T, \cdots, (C\Phi_N(s))^T]^T.$$

In [1, 43, 44], and many others, authors make the assumption that the function $f(\cdot)$ is in the QUAD function class. In this chapter, we make similar assumption:

(A1) There exist a positive definite diagonal matrix $Z = \mathrm{diag}(z_1, z_2, \cdots, z_n)$ and a diagonal matrix $\Delta = \mathrm{diag}(\delta_1, \delta_2, \cdots, \delta_n)$ such that $\hat{f}$ satisfies the following inequality:

$$(x - y)^T Z[\hat{f}(x) - \hat{f}(y) - \Delta(x - y)] \leqslant -\eta(x - y)^T(x - y),$$

for some $\eta > 0$ and all $x, y \in \mathbb{R}^n$.

For the convenience, we denote

$$\hat{Z} = \text{diag}(Z, Z, \cdots, Z),$$
$$\hat{\Delta} = \text{diag}(\Delta, \Delta, \cdots, \Delta).$$

Define the following Lyapunov functional for system (2.25)

$$V(t) = e^{\varepsilon t} e^T(t)\hat{Z}e(t) + \int_{t-\tau(t)}^{t} e^{\varepsilon s} e^T(s)Qe(s)ds.$$

By the similar proof of Theorems 2.3 and 2.4, we can obtain the following conclusions. Here we omit their proof to avoid the repetition.

Theorem 2.5 *Let (A1) hold, and $\dot{\tau}(t) \leqslant \sigma < 1$. If there exist matrix $Q > 0$ and constants $\varepsilon > 0$, $\mu > 0$, $M > 0$, $\eta_k \geqslant 0$, such that*

$$\begin{pmatrix} W_1 & b\hat{Z}[G \otimes (C\Gamma_2 C^{-1})] \\ b[G \otimes (C\Gamma_2 C^{-1})]^T \hat{Z} & -(1-\sigma)e^{-\varepsilon\tau} Q \end{pmatrix} \leqslant 0, \tag{2.26}$$

$$\mu\tau \leqslant \inf\{t_k - t_{k-1}, k \in \mathbb{N}\}, \tag{2.27}$$

$$\|I_k(e_i)\| \leqslant \eta_k \|e_i(t_k^-)\|, \tag{2.28}$$

$$\max\{\beta_k, k \in \mathbb{N}\} \leqslant M < e^{\varepsilon\mu\tau} \tag{2.29}$$

where $W_1 = \varepsilon\hat{Z} + a\hat{Z}(G \otimes (C\Gamma_1 C^{-1})) + a(G \otimes (C\Gamma_1 C^{-1}))^T \hat{Z} + Q + 2(-\eta I_{nN} + \hat{Z}\hat{\Delta})$, $\beta_k = 1 + \frac{(2\eta_k + \eta_k^2)\|\hat{Z}\|}{\lambda_m(\hat{Z})}$, $k \in \mathbb{N}$, $i = 1, 2, \cdots, N$, *then impulsive complex network* (2.1) *achieves global exponential output synchronization.*

Theorem 2.6 *Let (A1) hold, $\dot{\tau}(t) \leqslant \sigma < 1$ and $e_i(t_k) = r_{ik} e_i(t_k^-)$.*

(i) *If $r_k^2 \leqslant 1$ for all k, then impulsive complex network* (2.1) *achieves global exponential output synchronization if there exist matrix $Q > 0$ and a positive constant ε, such that*

$$\begin{pmatrix} W_1 & b\hat{Z}[G \otimes (C\Gamma_2 C^{-1})] \\ b[G \otimes (C\Gamma_2 C^{-1})]^T \hat{Z} & -(1-\sigma)e^{-\varepsilon\tau} Q \end{pmatrix} \leqslant 0 \tag{2.30}$$

where $i = 1, 2, \cdots, N$, $W_1 = \varepsilon\hat{Z} + a\hat{Z}(G \otimes (C\Gamma_1 C^{-1})) + a(G \otimes (C\Gamma_1 C^{-1}))^T \hat{Z} + Q + 2(-\eta I_{nN} + \hat{Z}\hat{\Delta})$, $r_k^2 = \max\{r_{ik}^2,\ i = 1, 2, \cdots, N\}$.

(ii) *If there exist k such that* $r_k^2 > 1$, *then impulsive complex network* (2.1) *achieves global exponential output synchronization if LMI* (2.30) *and the following conditions hold*

$$\mu\tau \leqslant \inf\{t_k - t_{k-1},\ k \in \mathbb{N}\},\ \mu > 0, \tag{2.31}$$

$$\max\{r_k^2, k \in \mathbb{N}\} \leqslant M < e^{\varepsilon\mu\tau}. \tag{2.32}$$

Remark 2.7 It should be noted that the proposed local and global exponential output synchronization criteria in Theorems 2.3, 2.4, 2.5, and 2.6 are expressed in terms of LMIs, which can be easily checked by LMI Toolbox in MATLAB. Therefore, our criteria are flexible and convenient.

2.4 Numerical Examples

In this section, we give two examples and their simulation to show the effectiveness of the above obtained theoretical results.

Example 2.8 Consider a complex dynamical network consisting of 10 identical nodes, in which each node is a 3-dimensional nonlinear system described by

$$\begin{pmatrix} \dot{x}_1 \\ \dot{x}_2 \\ \dot{x}_3 \end{pmatrix} = \begin{pmatrix} -6x_1 + 2x_2^2 \\ -7x_2 \\ -8x_3 + x_2x_3 \end{pmatrix}.$$

Take $\Gamma_1 = \text{diag}(0.5, 0.5, 0.5)$, $\Gamma_2 = \text{diag}(0.3, 0.3, 0.3)$, $C = \text{diag}(0.4, 0.8, 0.8)$, $a = 0.08$, $b = 0.06$, $\beta_\omega = 1$. Obviously, C is a nonsingular matrix. We can get

$$\dot{h}(t) = \begin{pmatrix} \dot{h}_1(t) \\ \dot{h}_2(t) \\ \dot{h}_3(t) \end{pmatrix} = \begin{pmatrix} -6h_1(t) + 1.25h_2^2(t) \\ -7h_2(t) \\ -8h_3(t) + 1.25h_2(t)h_3(t) \end{pmatrix}. \tag{2.33}$$

Clearly, $h(t) = (0, 0, 0)^T$ is an equilibrium point of the system (2.33), and the Jacobin matrix is

$$D\hat{f}(h(t)) = \begin{pmatrix} -6 & 0 & 0 \\ 0 & -7 & 0 \\ 0 & 0 & -8 \end{pmatrix}.$$

The matrix L is chosen as

$$L = \begin{pmatrix} -3 & 1 & 1 & 0 & 0 & 0 & 0 & 0 & 0 & 1 \\ 1 & -4 & 1 & 0 & 1 & 0 & 1 & 0 & 0 & 0 \\ 1 & 1 & -5 & 1 & 0 & 1 & 0 & 0 & 0 & 1 \\ 0 & 0 & 1 & -3 & 1 & 0 & 0 & 0 & 1 & 0 \\ 0 & 1 & 0 & 1 & -3 & 1 & 0 & 0 & 0 & 0 \\ 0 & 0 & 1 & 0 & 1 & -3 & 1 & 0 & 0 & 0 \\ 0 & 1 & 0 & 0 & 0 & 1 & -4 & 1 & 0 & 1 \\ 0 & 0 & 0 & 0 & 0 & 0 & 1 & -2 & 1 & 0 \\ 0 & 0 & 0 & 1 & 0 & 0 & 0 & 1 & -2 & 0 \\ 1 & 0 & 1 & 0 & 0 & 0 & 1 & 0 & 0 & -3 \end{pmatrix}.$$

It is obvious that such complex network is connected. Moreover, we can obtain

$$G = \begin{pmatrix} -1 & \frac{1}{3} & \frac{1}{3} & 0 & 0 & 0 & 0 & 0 & 0 & \frac{1}{3} \\ \frac{1}{4} & -1 & \frac{1}{4} & 0 & \frac{1}{4} & 0 & \frac{1}{4} & 0 & 0 & 0 \\ \frac{1}{5} & \frac{1}{5} & -1 & \frac{1}{5} & 0 & \frac{1}{5} & 0 & 0 & 0 & \frac{1}{5} \\ 0 & 0 & \frac{1}{3} & -1 & \frac{1}{3} & 0 & 0 & 0 & \frac{1}{3} & 0 \\ 0 & \frac{1}{3} & 0 & \frac{1}{3} & -1 & \frac{1}{3} & 0 & 0 & 0 & 0 \\ 0 & 0 & \frac{1}{3} & 0 & \frac{1}{3} & -1 & \frac{1}{3} & 0 & 0 & 0 \\ 0 & \frac{1}{4} & 0 & 0 & 0 & \frac{1}{4} & -1 & \frac{1}{4} & 0 & \frac{1}{4} \\ 0 & 0 & 0 & 0 & 0 & 0 & \frac{1}{2} & -1 & \frac{1}{2} & 0 \\ 0 & 0 & 0 & \frac{1}{2} & 0 & 0 & 0 & \frac{1}{2} & -1 & 0 \\ \frac{1}{3} & 0 & \frac{1}{3} & 0 & 0 & 0 & \frac{1}{3} & 0 & 0 & -1 \end{pmatrix},$$

which is weighted and asymmetric. We can find the matrix ϕ such that $\phi^{-1}G\phi = \Lambda = \text{diag}(0, -0.3764, -0.5757, -0.6968, -0.9427, -1.1635, -1.2721, -1.4823, -1.5626, -1.9279)$.

Since $h(t) = (0, 0, 0)^T$ is an equilibrium point of the system (2.33), then we can get $\nu_i(t_k) = r_k \nu_i(t_k^-)$ if $x_i(t_k) = r_k x_i(t_k^-)$, $k \in \mathbb{N}, i = 1, 2, \cdots, 10$. In the following, we analyze the local exponential output synchronization of impulsive complex network.

Case 1: set $r_{ik} = 0.5$, $t_k = 0.2k, k \in \mathbb{N}$, $\tau(t) = 0.5 - 0.5e^{-t}$. Then, we have $0 \leqslant \tau(t) \leqslant \tau = 0.5$, $\dot{\tau}(t) = 0.5e^{-t} \leqslant 0.5 < 1$, for $t \geqslant 0$, and $r_k \leqslant 1$ for all k. We can find the following positive definite matrices P and Q satisfying the LMI (2.19) with $\varepsilon = 0.2$,

$$P = \text{diag}(1.6754I_3, 1.8976I_3, 3.4321I_3, 2.4532I_3, 3.1254I_3, 4.2315I_3,$$
$$3.7865I_3, 4.3246I_3, 3.5643I_3, 2.4563I_3),$$
$$Q = \text{diag}(4.3754I_3, 3.5643I_3, 4.8921I_3, 3.2685I_3, 4.9867I_3, 4.9675I_3,$$
$$5.9765I_3, 4.8954I_3, 6.5689I_3, 5.6763I_3).$$

According to Theorem 2.4, the impulsive complex network achieves local exponential output synchronization. The simulation results are shown in Fig. 2.1.

Case 2: set $r_{ik} = 1 + \frac{1}{10+t_k}$, $t_k = k, k \in \mathbb{N}, \tau(t) = 0.5 - 0.5e^{-t}$. Then, we have $0 \leqslant \tau(t) \leqslant \tau = 0.5$, $\dot{\tau}(t) = 0.5e^{-t} \leqslant 0.5 < 1$, for $t \geqslant 0$, and $r_k > 1$ for all k.

We can find the following positive definite matrices P and Q satisfying (2.19), (2.20), and (2.21) with $\varepsilon = 0.2$, $\mu = 2$, $M = 1.21$,

$$P = \text{diag}(1.6754I_3, 1.8976I_3, 3.4321I_3, 2.4532I_3, 3.1254I_3, 4.2315I_3,$$
$$3.7865I_3, 4.3246I_3, 3.5643I_3, 2.4563I_3),$$
$$Q = \text{diag}(4.3754I_3, 3.5643I_3, 4.8921I_3, 3.2685I_3, 4.9867I_3, 4.9675I_3,$$
$$5.9765I_3, 4.8954I_3, 6.5689I_3, 5.6763I_3).$$

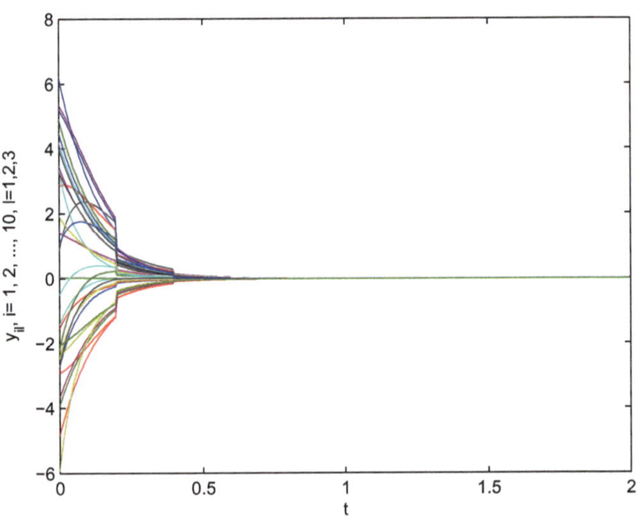

Fig. 2.1 Output synchronization of impulsive complex network with time-varying delay in Case 1

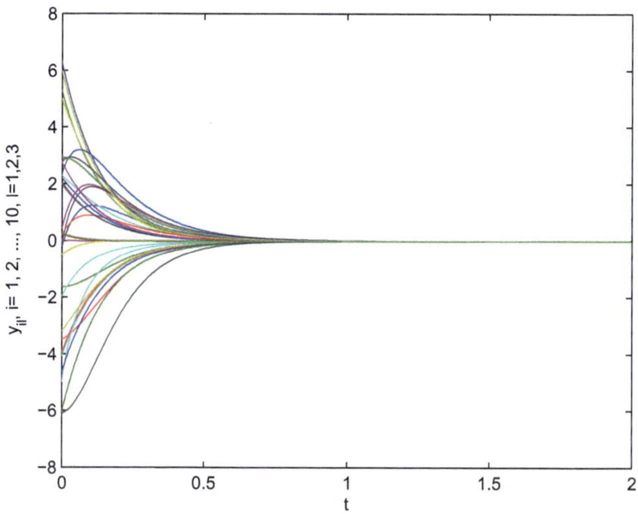

Fig. 2.2 Output synchronization of impulsive complex network with time-varying delay in Case 2

It follows from Theorem 2.4 that the impulsive complex network achieves local exponential output synchronization. The simulation results are shown in Fig. 2.2.

According to the change processes of the output variables in Figs. 2.1 and 2.2, we clearly see that the outputs are synchronization.

Next, we analyze the global exponential output synchronization of impulsive complex network.

Example 2.9 Consider a complex dynamical network consisting of 10 identical nodes, in which each node is a 3-dimensional nonlinear system described by

$$\begin{pmatrix} \dot{x}_1 \\ \dot{x}_2 \\ \dot{x}_3 \end{pmatrix} = \begin{pmatrix} 10(x_2 - x_1) \\ 2x_1 - x_1x_3 - 10x_2 \\ x_1x_2 - \frac{8}{3}x_3 \end{pmatrix}.$$

Take $\Gamma_1 = \text{diag}(0.6, 0.6, 0.6)$, $\Gamma_2 = \text{diag}(0.4, 0.4, 0.4)$, $a = 0.05$, $b = 0.04$, $\beta_\omega = 1$, $C = \text{diag}(0.5, 0.5, 0.5)$. Obviously, C is a nonsingular matrix. We can obtain

$$\dot{h}(t) = \begin{pmatrix} \dot{h}_1(t) \\ \dot{h}_2(t) \\ \dot{h}_3(t) \end{pmatrix} = \begin{pmatrix} 10(h_2(t) - h_1(t)) \\ 2h_1(t) - 2h_1(t)h_3(t) - 10h_2(t) \\ 2h_1(t)h_2(t) - \frac{8}{3}h_3(t) \end{pmatrix}. \tag{2.34}$$

Clearly, $h(t) = (0, 0, 0)^T$ is an equilibrium point of the system (2.34) and $\hat{f}(0) = 0$. In this case, if $\hat{f}$ satisfies the following inequality:

$$x^T Z(\hat{f}(x) - \Delta x) \leqslant -\eta x^T x$$

for all $x \in \mathbb{R}^n$, where $\eta > 0$, $Z = \text{diag}(z_1, z_2, \cdots, z_n)$, $\Delta = \text{diag}(\delta_1, \delta_2, \cdots, \delta_n)$, $z_i > 0, i = 1, 2, \cdots, n$, then we can obtain Theorems 2.5 and 2.6. Thus, we can take $\eta = \frac{8}{3}$, $Z = \text{diag}(1, 1, 1)$, $\Delta = \text{diag}(0, 0, 0)$.

The matrix L is chosen as

$$L = \begin{pmatrix} -2 & 0 & 0 & 0 & 0 & 0 & 0 & 0 & 1 & 1 \\ 0 & -1 & 1 & 0 & 0 & 0 & 0 & 0 & 0 & 0 \\ 0 & 1 & -2 & 0 & 0 & 0 & 0 & 0 & 0 & 1 \\ 0 & 0 & 0 & -1 & 0 & 0 & 0 & 0 & 0 & 1 \\ 0 & 0 & 0 & 0 & -1 & 0 & 0 & 0 & 0 & 1 \\ 0 & 0 & 0 & 0 & 0 & -1 & 0 & 0 & 1 & 0 \\ 0 & 0 & 0 & 0 & 0 & 0 & -1 & 1 & 0 & 0 \\ 0 & 0 & 0 & 0 & 0 & 0 & 1 & -2 & 1 & 0 \\ 1 & 0 & 0 & 0 & 0 & 1 & 0 & 1 & -4 & 1 \\ 1 & 0 & 1 & 1 & 1 & 0 & 0 & 0 & 1 & -5 \end{pmatrix}.$$

It is obvious that such complex network is connected. Moreover, we can get

$$G = \begin{pmatrix} -1 & 0 & 0 & 0 & 0 & 0 & 0 & 0 & \frac{1}{2} & \frac{1}{2} \\ 0 & -1 & 1 & 0 & 0 & 0 & 0 & 0 & 0 & 0 \\ 0 & \frac{1}{2} & -1 & 0 & 0 & 0 & 0 & 0 & 0 & \frac{1}{2} \\ 0 & 0 & 0 & -1 & 0 & 0 & 0 & 0 & 0 & 1 \\ 0 & 0 & 0 & 0 & -1 & 0 & 0 & 0 & 0 & 1 \\ 0 & 0 & 0 & 0 & 0 & -1 & 0 & 0 & 1 & 0 \\ 0 & 0 & 0 & 0 & 0 & 0 & -1 & 1 & 0 & 0 \\ 0 & 0 & 0 & 0 & 0 & 0 & \frac{1}{2} & -1 & \frac{1}{2} & 0 \\ \frac{1}{4} & 0 & 0 & 0 & 0 & \frac{1}{4} & 0 & \frac{1}{4} & -1 & \frac{1}{4} \\ \frac{1}{5} & 0 & \frac{1}{5} & \frac{1}{5} & \frac{1}{5} & 0 & 0 & 0 & \frac{1}{5} & -1 \end{pmatrix},$$

which is weighted and asymmetric.

Since $h(t) = (0, 0, 0)^T$ is an equilibrium point of the system (2.34), then we can get $e_i(t_k) = r_{ik} e_i(t_k^-)$ if $x_i(t_k) = r_{ik} x_i(t_k^-)$, $k \in \mathbb{N}, i = 1, 2, \cdots, 10$.

Case 1: set $r_{ik} = \frac{2}{3}$, $t_k = 0.4k$, $k \in \mathbb{N}, i = 1, 2, \cdots, 10$, $\tau(t) = 0.4 - 0.4e^{-t}$. Then, we have $0 \leqslant \tau(t) \leqslant \tau = 0.4$, $\dot{\tau}(t) = 0.4e^{-t} \leqslant 0.4 < 1$, for $t \geqslant 0$, and $r_k \leqslant 1$ for all k.

We can find the following positive definite matrix Q satisfying the LMI (2.30) with $\varepsilon = 0.2$,

$$Q = \text{diag}(2.4378 I_3, 3.5943 I_3, 3.5437 I_3, 2.9807 I_3, 3.4532 I_3, 2.8769 I_3, 3.7548 I_3, 2.5748 I_3, 3.3896 I_3, 4.2685 I_3).$$

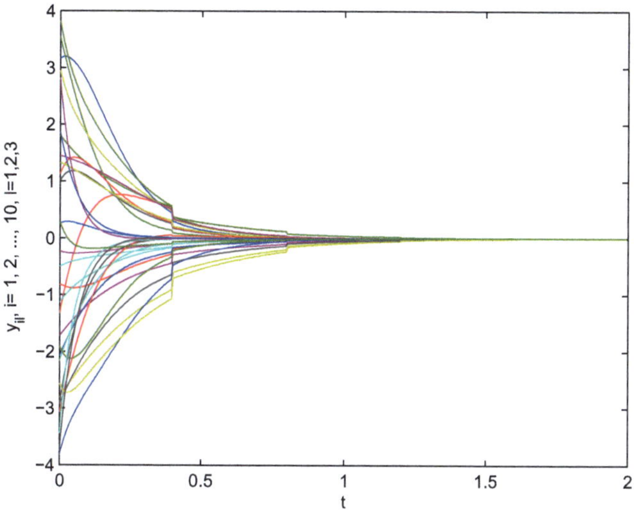

Fig. 2.3 Output synchronization of impulsive complex network with time-varying delay in Case 1

According to Theorem 2.6, the impulsive complex network achieves global exponential output synchronization. The simulation results are shown in Fig. 2.3.

Case 2: set $r_{ik} = 1 + \frac{1}{9+i+t_k}$, $t_k = k, k \in \mathbb{N}, i = 1, 2, \cdots, 10, \tau(t) = 0.4 - 0.4e^{-t}$. Then, we have $0 \leqslant \tau(t) \leqslant \tau = 0.4$, $\dot{\tau}(t) = 0.4e^{-t} \leqslant 0.4 < 1$, for $t \geqslant 0$, and $r_k > 1$ for all k.

We can find the following positive definite matrix Q satisfying (2.30), (2.31), and (2.32) with $\varepsilon = 0.2$, $\mu = 2.5$, $M = 1.2$,

$$Q = \text{diag}(2.4378I_3, 3.5943I_3, 3.5437I_3, 2.9807I_3, 3.4532I_3, 2.8769I_3, \\ 3.7548I_3, 2.5748I_3, 3.3896I_3, 4.2685I_3).$$

It follows from Theorem 2.6 that the impulsive complex network achieves global exponential output synchronization. The simulation results are shown in Fig. 2.4.

According to the change processes of the output variables in Figs. 2.3 and 2.4, we clearly see that the outputs are synchronization.

Remark 2.10 In recent years, some researchers have studied the synchronization of complex dynamical networks [67, 68]. It should be noticed that the state synchronization of complex networks without time delays was considered in [67, 68]. It is well known that the time delays always exist in complex networks due to the finite speeds of transmission and/or the traffic congestion, and most of delays is notable. So it is crucial for us to take the delay into the consideration when we study complex networks. Therefore, it is interesting to study the output synchronization of impulsive complex networks with time-varying delay.

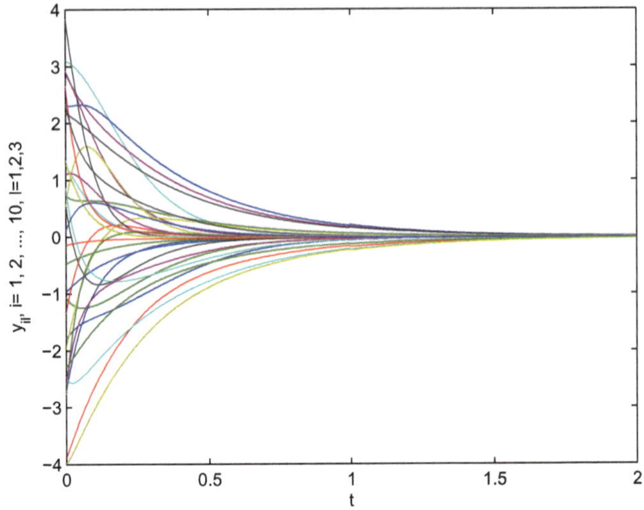

Fig. 2.4 Output synchronization of impulsive complex network with time-varying delay in Case 2

2.5 Conclusion

A new impulsive complex network model has been introduced. Both local and global exponential output synchronization have been taken into consideration in this chapter, and some sufficient conditions have been derived. Numerical simulations have been provided to verify the proposed results. In future work we shall study the local and global exponential output synchronization of impulsive complex networks, in which the output matrix C is singular.

Chapter 3
Passivity and Output Synchronization of CDNs with Fixed and Adaptive Coupling Strength

3.1 Introduction

In the real world, many systems can be modeled as complex networks. Examples include food webs, communication networks, social networks, power grids, cellular networks, World Wide Web, metabolic systems, disease transmission networks, and so on. Therefore, the analysis and control of dynamical behaviors in CDNs have received considerable attention in recent years.

Passivity is part of a broader and a general theory of dissipativeness [42, 69]. The main point of passivity theory is that the passive properties of systems can keep the systems internally stable. The passivity theory was firstly proposed in the circuit analysis [70] and since then has found successful applications in diverse areas such as stability [71, 72], complexity [73], signal processing [74], chaos control and synchronization [75, 76], fuzzy control [77], and so on. These are the main reasons why passivity theory has become a very hot topic across many fields, and much investigative attention has been focused on this topic. Recently, many authors have studied the passivity of CDNs, and a great many important results on passivity have been obtained for various complex networks such as time invariant, time-varying, discrete, uncertain, impulsive, and stochastic network models; see [1, 60, 61, 78–84] and relevant references therein. For instance, Yao et al. [61] considered a class of complex spatio-temporal switching networks with time delays, and several passivity criteria were established by constructing appropriate Lyapunov-Krasovskii functionals. In [79], the problem of passivity analysis was studied for discrete-time stochastic Markovian jump neural networks with both discrete and distributed delays. Wang, Yang and Wu [1] discussed a complex network model with non-linear, time-varying, non-symmetric and delayed coupling, and some sufficient conditions on input passivity and output passivity were obtained by Lyapunov functional method. Although research on passivity has attracted so much attention, little of that had been devoted to the passivity properties of the spatially and temporally CDNs until Wang, Wu and Guo [82] obtained the

J.-L. Wang et al., *Analysis and Control of Output Synchronization for Complex Dynamical Networks*, https://doi.org/10.1007/978-981-13-1352-3_3

conditions for passivity of reaction-diffusion neural networks. In [83], the robust passivity was investigated for a class of parabolic complex networks with spatially and temporally varying input and output variables. Wang and Wu [84] analyzed the passivity of a class of delayed reaction-diffusion networks by using the Lyapunov functional method, and also revealed the relationship between passivity and stability of the delayed reaction-diffusion networks.

Unfortunately, in most existing works on the passivity of complex networks (see also the above mentioned references), they always assume that the node input has the same dimension as the output vector. But, in reality, the dimensions of the input and output are different in many circumstances. Therefore, it is interesting to further study the passivity of CDNs, in which the input and output have different dimensions. On the other hand, in many real-world networks, the coupling strength is adaptively adjusted according to changes in the environment or the network itself (for instance, neural networks, wireless sensor networks, biological networks [85]). To our knowledge, the passivity of complex networks with adaptive coupling strength and different dimensions of input and output has not yet been investigated. Hence, the passivity problem of complex networks with adaptive coupling strength is also considered in this paper.

As one of the most significant and interesting dynamical properties of the complex networks [86, 87], synchronization has received much of the focus in recent years. A wide variety of synchronization criteria have been presented for various complex networks [6, 8, 9, 16, 55, 56, 60, 61, 88–92]. In [55], the authors studied the synchronization problem of a coupled reaction-diffusion neural networks with directed topologies by utilizing the pinning control technique and the Lyapunov functional method. Liu, Su and Chen [91] investigated finite-time synchronization of a CNNs with the fixed network topology and control strength. In [92], Yang, Cao and Yang considered the global exponential synchronization of coupled reaction-diffusion neural networks with time-varying delays by using pinning-impulsive control strategy. The passivity theory has long been a nice tool for analyzing the synchronization of the complex networks [33, 35, 60, 61, 93–95]. Based on the passivity property, some sufficient conditions for synchronization of the CDNs with and without coupling delays were derived in [60]. In [33, 35, 93], the authors studied the output synchronization problem for CDNs with nonidentical nodes by using the passivity property. By exploiting passivity-based synchronization results, an adaptive control algorithm was proposed to guarantee task-space synchronization of networked robotic manipulators in the presence of dynamic uncertainties and time-varying communication delays [94]. In [95], the authors presented some sufficient conditions for synchronization in networks of diffusively coupled neuronal oscillators using the theory of semi-passive and passive systems. To our knowledge, in most existing works on this topic (see also the above mentioned references), it is assumed that the input vector has the same dimension as the output vector. It is well known that the dimensions of the input and output vectors are different in many circumstances. Therefore, in this chapter, we also use passivity theory to study the output synchronization problem of CDNs, in which the input and output vectors have different dimensions.

The main contributions of this chapter are as follows. First, some criteria for passivity, input strict passivity and output strict passivity of the complex dynamical network with fixed coupling strength are derived. In addition, we reveal the relationship between output strict passivity and output synchronization, and establish a criterion for output synchronization by employing the relationship between output strict passivity and output synchronization. Second, we respectively investigate the passivity, input strict passivity and output strict passivity of the complex dynamical network with adaptive coupling strength, and reveal the relationship between output strict passivity and output synchronization. By utilizing the designed adaptive law and the relationship between output strict passivity and output synchronization, a sufficient condition is obtained to guarantee the output synchronization of the complex dynamical network.

The rest of this chapter is organized as follows. In Sect. 3.2, our mathematical model of complex dynamical network is presented and some preliminaries are given. The main results of this chapter are given in Sects. 3.3 and 3.4. In Sect. 3.5, two numerical examples are provided to illustrate the effectiveness of the proposed results. Finally, Sect. 3.6 concludes the investigation.

3.2 Network Model

In this chapter, we consider a complex network consisting of N linearly and diffusively coupled identical nodes, in which each node is an n-dimensional dynamical system. The mathematical model of the complex network can be described as follows:

$$\begin{cases} \dot{x}_i(t) = f(x_i(t)) + c\sum_{j=1}^{N} G_{ij}\Gamma x_j(t) + Bu_i(t), \\ y_i(t) = \quad Cx_i(t) + Du_i(t), \quad t \geqslant 0, \end{cases} \tag{3.1}$$

where $i = 1, 2, \cdots, N$, N is the number of nodes in the network, $f(\cdot) \in \mathbb{R}^n$ is a continuously differentiable vector function describing the dynamics of each individual node, $x_i(t) = (x_{i1}(t), x_{i2}(t), \cdots, x_{in}(t))^T \in \mathbb{R}^n$ is the state vector of node i, $u_i(t) = (u_{i1}(t), u_{i2}(t), \cdots, u_{ip}(t))^T \in \mathbb{R}^p$ and $y_i(t) = (y_{i1}(t), y_{i2}(t), \cdots, y_{iq}(t))^T \in \mathbb{R}^q$ denote input and output of node i, respectively, B, C and D are known matrices with appropriate dimensions, c is a positive real number, which represents the overall coupling strength, $\Gamma \in \mathbb{R}^{n\times n}$ is the inner coupling matrix, $G = (G_{ij})_{N\times N}$ represents the topological structure of the network, where G_{ij} is defined as follows: if there exists a connection between node i and node j, then $G_{ij} = G_{ji} > 0$, otherwise, $G_{ij} = G_{ji} = 0 (i \neq j)$, and the diagonal elements of matrix G are defined by

$$G_{ii} = -\sum_{\substack{j=1 \\ j\neq i}}^{N} G_{ij}, \quad i = 1, 2, \cdots, N.$$

Remark 3.1 The passivity problem of CDNs has received much of the focus in recent years. So far, a great many important results on passivity have been obtained for various complex networks; see [1, 60, 61, 78–84] and relevant references therein. Unfortunately, in most existing works on the passivity of complex networks, it is assumed that the input vector has the same dimension as the output vector. Practically, the dimensions of the input and output vectors are different in many circumstances. Therefore, it is important and interesting to further study the passivity of CDNs, in which the input and output vectors have different dimensions.

Remark 3.2 It is well known that the input and output vectors in many systems have different dimensions. However, this case can not be dealt with by traditional passivity theory. As a natural extension of the existing concepts of passivity [1, 60, 61, 81, 82, 84], we propose two new passivity definitions (see Definitions 1.2 and 1.3) for the case where the dimensions of the input and output vectors are different. Letting $P = I_p$, $W = \gamma_1 I_p$ and $Q = \gamma_2 I_p$, then these traditional passivity definitions in [1, 60, 61, 81, 82, 84] can be obtained by Definitions 1.2 and 1.3.

Definition 3.3 The complex network (3.1) is said to achieve output synchronization if

$$\lim_{t \to +\infty} ||y_i(t) - y_j(t)|| = 0 \text{ for all } i, j = 1, 2, \cdots, N$$

under the condition that $u_i(t) = 0$, $i = 1, 2, ..., N$.

Remark 3.4 In order to guarantee the existence and uniqueness of the solutions of complex network (3.1), we always assume function $f(\cdot)$ satisfies global Lipschitz condition in this chapter. The Lemma 1.8 (Barbalat's Lemma) will play an important role in the synchronization analysis of the output-strictly passive CDNs.

3.3 Passivity and Output Synchronization of CDNs with Fixed Coupling Strength

In this section, we analyze the passivity, input strict passivity and output strict passivity of complex network (3.1), and reveal the relationship between output strict passivity and output synchronization. Moreover, by using the relationship between output strict passivity and output synchronization, a criterion is obtained to guarantee the output synchronization of the complex network (3.1).

3.3.1 Passivity Analysis of CDNs

In order to obtain our main results, an assumption is introduced in this subsection.

(A1) There exist a positive definite diagonal matrix $M = \text{diag}(m_1, m_2, \cdots, m_n) \in \mathbb{R}^{n \times n}$ and a diagonal matrix $\Delta = \text{diag}(\delta_1, \delta_2, \cdots, \delta_n) \in \mathbb{R}^{n \times n}$ such that $f(\cdot)$ satisfies the following inequality:

$$x^T M(f(x) - \Delta x) \leqslant -\eta x^T x$$

for some $\eta \in \mathbb{R} > 0$ and all $x \in \mathbb{R}^n$.

Theorem 3.5 *Let (A1) hold. If there exists a matrix $P \in \mathbb{R}^{qN\times pN}$ such that*

$$\begin{pmatrix} \mathbb{W}_1 & \mathbb{H}_1 \\ \mathbb{H}_1^T & -\frac{(I_N\otimes D^T)P+P^T(I_N\otimes D)}{2} \end{pmatrix} \leqslant 0, \tag{3.2}$$

where $\mathbb{W}_1 = 2[I_N \otimes (M\Delta - \eta I_n)] + c[G \otimes (M\Gamma + \Gamma^T M)]$, $\mathbb{H}_1 = I_N \otimes (MB) - \frac{(I_N\otimes C^T)P}{2}$, *then the complex network* (3.1) *is passive in the sense of Definition* 1.2.

Proof First, construct a Lyapunov functional for system (3.1) as follows:

$$V_1(t) = \sum_{i=1}^{N} x_i^T(t)Mx_i(t). \tag{3.3}$$

Calculating the derivative of $V_1(t)$ along the trajectory of system (3.1), we can get

$$\begin{aligned}
\dot{V}_1(t) &= 2\sum_{i=1}^{N} x_i^T(t)M\dot{x}_i(t) \\
&= 2\sum_{i=1}^{N} x_i^T(t)M[f(x_i(t)) + c\sum_{j=1}^{N} G_{ij}\Gamma x_j(t) + Bu_i(t)] \\
&\leqslant 2\sum_{i=1}^{N} x_i^T(t)(M\Delta - \eta I_n)x_i(t) + 2\sum_{i=1}^{N} x_i^T(t)MBu_i(t) \\
&\quad +2c\sum_{i=1}^{N}\sum_{j=1}^{N} G_{ij}x_i^T(t)M\Gamma x_j(t).
\end{aligned}$$

Let $x(t) = (x_1^T(t), x_2^T(t), \cdots, x_N^T(t))^T$, $u(t) = (u_1^T(t), u_2^T(t), \cdots, u_N^T(t))^T$ and $y(t) = (y_1^T(t), y_2^T(t), \cdots, y_N^T(t))^T$. Then, we have

$$\begin{aligned}
&\dot{V}_1(t) - y^T(t)Pu(t) \\
&\leqslant 2\sum_{i=1}^{N} x_i^T(t)(M\Delta - \eta I_n)x_i(t) + 2\sum_{i=1}^{N} x_i^T(t)MBu_i(t) \\
&\quad +2c\sum_{i=1}^{N}\sum_{j=1}^{N} G_{ij}x_i^T(t)M\Gamma x_j(t)
\end{aligned}$$

$$
\begin{aligned}
&- [(I_N \otimes C)x(t) + (I_N \otimes D)u(t)]^T Pu(t) \\
&= \xi^T(t) \begin{pmatrix} \mathbb{W}_1 & \mathbb{H}_1 \\ \mathbb{H}_1^T & -\frac{(I_N \otimes D^T)P + P^T(I_N \otimes D)}{2} \end{pmatrix} \xi(t),
\end{aligned}
$$

where $\xi(t) = (x^T(t), u^T(t))^T$.

From (3.2), we have

$$
\dot{V}_1(t) \leqslant y^T(t)Pu(t). \tag{3.4}
$$

By integrating equation (3.4) with respect to t over the time period t_0 to t_p, we can obtain

$$
\int_{t_0}^{t_p} y^T(t)Pu(t)dt \geqslant V_1(t_p) - V_1(t_0)
$$

for any $t_p, t_0 \in \mathbb{R}^+$ and $t_p \geqslant t_0$. The proof is completed.

Remark 3.6 In [96], Hill and Moylan investigated a class of nonlinear ordinary differential equation systems. Under the assumption that system is passive and zero-state detectable, the authors proved that system is stable. Wang and Wu [84] analyzed the passivity of a class of reaction-diffusion networks, in which the input and output variables are varied with the time and space variables. In addition, they proved that the reaction-diffusion network is stable if it is passive with respect to a class of storage functions. Practically, these results can also be extended to the case when the input and output vectors have different dimensions. Similar to the proof of Theorem 3.11 in [84], we can easily derive that complex network (3.1) is stable if it is passive with respect to a positive definite storage function (e.g., $V_1(t)$) and $f(0) = 0$.

By the similar proof of Theorem 3.5, we can obtain the following conclusions. Here we omit their proof to avoid the repetition.

Theorem 3.7 *Let (A1) hold. If there exist a matrix $W \in \mathbb{R}^{pN \times pN} > 0$, and a matrix $P \in \mathbb{R}^{qN \times pN}$ such that*

$$
\begin{pmatrix} \mathbb{W}_1 & \mathbb{H}_1 \\ \mathbb{H}_1^T & W - \frac{(I_N \otimes D^T)P + P^T(I_N \otimes D)}{2} \end{pmatrix} \leqslant 0, \tag{3.5}
$$

where $\mathbb{W}_1 = 2[I_N \otimes (M\Delta - \eta I_n)] + c[G \otimes (M\Gamma + \Gamma^T M)]$, $\mathbb{H}_1 = I_N \otimes (MB) - \frac{(I_N \otimes C^T)P}{2}$, then the complex network (3.1) *is input-strictly passive in the sense of Definition* 1.3.

Remark 3.8 Similarly, if complex network (3.1) is input-strictly passive with respect to a positive definite storage function and $f(0) = 0$, we can easily obtain that complex network (3.1) is stable. This can be viewed as the extension of the results in [84, 96].

Theorem 3.9 *Let (A1) hold. If there exist a matrix* $Q \in \mathbb{R}^{qN \times qN} > 0$*, and a matrix* $P \in \mathbb{R}^{qN \times pN}$ *such that*

$$\begin{pmatrix} \mathbb{W}_2 & \mathbb{W}_3 \\ \mathbb{W}_3^T & \mathbb{H}_2 \end{pmatrix} \leqslant 0, \tag{3.6}$$

where $\mathbb{W}_2 = 2[I_N \otimes (M\Delta - \eta I_n)] + c[G \otimes (M\Gamma + \Gamma^T M)] + (I_N \otimes C^T)Q(I_N \otimes C)$, $\mathbb{W}_3 = I_N \otimes (MB) - \frac{(I_N \otimes C^T)P}{2} + (I_N \otimes C^T)Q(I_N \otimes D)$, $\mathbb{H}_2 = (I_N \otimes D^T)Q(I_N \otimes D) - \frac{(I_N \otimes D^T)P + P^T(I_N \otimes D)}{2}$, *then the complex network* (3.1) *is output-strictly passive in the sense of Definition* 1.3.

Remark 3.10 By constructing appropriate Lyapunov functional, some sufficient conditions ensuring the passivity, input strict passivity and output strict passivity are established for the complex network (3.1) (see Theorems 3.5, 3.7, 3.9). To our knowledge, this is the first time to investigate the passivity of CDNs with different dimensions of input and output, which is a very important problem. We generalize the existing passivity results to some extent, and this can serve as a stepping stone to study the passivity of complex networks, in which the input and output vectors have different dimensions.

3.3.2 Passivity-Based Output Synchronization of CDNs

The passivity theory has long been a nice tool for analyzing the output synchronization of complex networks. But in most existing works, it is assumed that the input vector has the same dimension as the output vector. Therefore, it is important and interesting to study the relationship between passivity and output synchronization of CDNs with different dimensions of input and output.

Let $\bar{x}(t) = \frac{1}{N}\sum_{i=1}^{N} x_i(t)$. Then, we have

$$\dot{\bar{x}}(t) = \frac{1}{N}\sum_{i=1}^{N} f(x_i(t)) + \frac{1}{N}\sum_{i=1}^{N} Bu_i(t).$$

Defining $e_i(t) = (e_{i1}(t), e_{i2}(t), \cdots, e_{in}(t))^T = x_i(t) - \bar{x}(t)$, $i = 1, 2, \cdots, N$, then the dynamics of $e_i(t)$ is governed by the following equation:

$$\begin{aligned} \dot{e}_i(t) = {} & f(x_i(t)) - \frac{1}{N}\sum_{i=1}^{N} f(x_i(t)) + c\sum_{j=1}^{N} G_{ij}\Gamma e_j(t) \\ & + Bu_i(t) - \frac{1}{N}\sum_{i=1}^{N} Bu_i(t), \end{aligned} \tag{3.7}$$

where $i = 1, 2, \cdots, N$.

On the other hand, there obviously exist nonsingular matrices $U \in \mathbb{R}^{q\times q}$ and $H \in \mathbb{R}^{n\times n}$ such that

$$C = U \begin{pmatrix} I_r & 0 \\ 0 & 0 \end{pmatrix} H,$$

where $r = \text{Rank}(C)$. Then, we can get

$$\begin{cases} \dot{e}_i(t) = & f(x_i(t)) - \frac{1}{N}\sum_{i=1}^{N} f(x_i(t)) + Bu_i(t) \\ & +c\sum_{j=1}^{N} G_{ij}\Gamma e_j(t) - \frac{1}{N}\sum_{i=1}^{N} Bu_i(t), \\ y_i(t) = & U\begin{pmatrix} I_r & 0 \\ 0 & 0 \end{pmatrix} He_i(t) + C\bar{x}(t) + Du_i(t), \end{cases}$$

where $i = 1, 2, \cdots, N$.

Obviously, the complex network (3.1) achieves output synchronization if and only if $\hat{y}_i(t), i = 1, 2, \cdots, N$, in the following system asymptotically converge to zero under the condition that $u_i(t) = 0$:

$$\begin{cases} \dot{e}_i(t) = & f(x_i(t)) - \frac{1}{N}\sum_{i=1}^{N} f(x_i(t)) + Bu_i(t) \\ & +c\sum_{j=1}^{N} G_{ij}\Gamma e_j(t) - \frac{1}{N}\sum_{i=1}^{N} Bu_i(t), \\ \hat{y}_i(t) = & \hat{H}e_i(t) + \hat{D}u_i(t), \end{cases} \tag{3.8}$$

where $\hat{y}_i(t) = (\hat{y}_{i1}(t), \hat{y}_{i2}(t), \cdots, \hat{y}_{ir}(t))^T \in \mathbb{R}^r$, $\hat{H} = (I_r \quad 0)_{r\times n}H$, $\hat{D} = (I_r \quad 0)_{r\times q}D$, $i = 1, 2, \cdots, N$.

For the convenience, we denote

$$u(t) = (u_1^T(t), u_2^T(t), \cdots, u_N^T(t))^T,$$
$$\hat{y}(t) = (\hat{y}_1^T(t), \hat{y}_2^T(t), \cdots, \hat{y}_N^T(t))^T,$$
$$e(t) = (e_1^T(t), e_2^T(t), \cdots, e_N^T(t))^T.$$

Theorem 3.11 *Suppose that* $V : \mathbb{R}^{nN} \to \mathbb{R}^+$ *is continuously differentiable and satisfies the following condition*

$$v_1(\|e(t)\|) \leqslant V(e(t)), \tag{3.9}$$

where $v_1 : \mathbb{R}^+ \to \mathbb{R}^+$ *is a strictly increasing continuous function and* $v_1(s) \to +\infty$ *as* $s \to +\infty$. *The complex network* (3.1) *achieves output synchronization if system* (3.8) *is output-strictly passive with respect to storage function* $V(e(t))$.

Proof If system (3.8) is output-strictly passive with respect to storage function $V(e(t))$, then there exist a matrix $Q \in \mathbb{R}^{rN\times rN} > 0$ and a matrix $P \in \mathbb{R}^{rN\times pN}$ such that

$$\int_t^{t+\varepsilon} (\hat{y}^T(s)Pu(s) - \hat{y}^T(s)Q\hat{y}(s))ds \geqslant V(e(t+\varepsilon)) - V(e(t)) \tag{3.10}$$

for any $t \in \mathbb{R}^+$ and $\varepsilon > 0$.

From (3.10), we can obtain

$$\frac{V(e(t+\varepsilon)) - V(e(t))}{\varepsilon} \leqslant \frac{\int_t^{t+\varepsilon} \hat{y}^T(s)Pu(s)ds}{\varepsilon} - \frac{\int_t^{t+\varepsilon} \hat{y}^T(s)Q\hat{y}(s)ds}{\varepsilon}. \tag{3.11}$$

By taking limit $\epsilon \to 0$ in (3.11), we have

$$\dot{V}(e(t)) \leqslant \hat{y}^T(t)Pu(t) - \hat{y}^T(t)Q\hat{y}(t). \tag{3.12}$$

Letting $u_i(t) = 0, i = 1, 2, \cdots, N$, we can get from (3.12) that

$$\begin{aligned}\dot{V}(e(t)) &\leqslant -e^T(t)(I_N \otimes \hat{H}^T)Q(I_N \otimes \hat{H})e(t) \\ &\leqslant -\lambda_m(Q)\sum_{i=1}^{N} e_i^T(t)\hat{H}^T\hat{H}e_i(t).\end{aligned} \tag{3.13}$$

Obviously, $V(e(t))$ is nonincreasing and $e(t)$ is bounded. Since $V(e(t)) \geqslant 0$ (see (3.9)), one can conclude that $\lim_{t\to+\infty} V(e(t))$ exists and is a nonnegative real number. Namely,

$$\lim_{t\to+\infty} V(e(t)) = \sigma \in \mathbb{R} \geqslant 0. \tag{3.14}$$

Let $\beta(t) = \sum_{i=1}^{N} e_i^T(t)\hat{H}^T\hat{H}e_i(t)$. From (3.13) and (3.14), one has

$$\begin{aligned}\lim_{t\to+\infty}\int_0^t \beta(s)ds &\leqslant -\frac{1}{\lambda_m(Q)}\lim_{t\to+\infty}\int_0^t \dot{V}(e(s))ds \\ &= \frac{1}{\lambda_m(Q)}\Big(V(e(0)) - \lim_{t\to+\infty} V(e(t))\Big) \\ &= \frac{1}{\lambda_m(Q)}\Big(V(e(0)) - \sigma\Big).\end{aligned} \tag{3.15}$$

Furthermore,

$$
\begin{aligned}
\dot{\beta}(t) &= 2\sum_{i=1}^{N} e_i^T(t)\hat{H}^T\hat{H}\dot{e}_i(t) \\
&= 2\sum_{i=1}^{N} e_i^T(t)\hat{H}^T\hat{H}\Big[f(x_i(t)) - f(\bar{x}(t)) + f(\bar{x}(t)) \\
&\quad -\frac{1}{N}\sum_{i=1}^{N} f(x_i(t)) + c\sum_{j=1}^{N} G_{ij}\Gamma e_j(t)\Big] \\
&= 2\sum_{i=1}^{N} e_i^T(t)\hat{H}^T\hat{H}\Big[f(x_i(t)) - f(\bar{x}(t)) \\
&\quad +c\sum_{j=1}^{N} G_{ij}\Gamma e_j(t)\Big].
\end{aligned}
\tag{3.16}
$$

Since $e_i(t), i = 1, 2, \cdots, N$, are bounded, we have

$$
\|e_i(t)\| \leqslant \lambda \in \mathbb{R} > 0 \tag{3.17}
$$

for any $t \in [0, +\infty)$.

Because $f(\cdot)$ satisfies global Lipschitz condition (see Remark 3.4), there exists a positive constant L such that

$$
\|f(x_i(t)) - f(\bar{x}(t))\| \leqslant L\|x_i(t) - \bar{x}(t)\| = L\|e_i(t)\|, \tag{3.18}
$$

where $i = 1, 2, \cdots, N$.

By (3.16), (3.17), and (3.18), one obtains

$$
\begin{aligned}
|\dot{\beta}(t)| &= \Big|2\sum_{i=1}^{N} e_i^T(t)\hat{H}^T\hat{H}\Big[f(x_i(t)) - f(\bar{x}(t)) \\
&\quad +c\sum_{j=1}^{N} G_{ij}\Gamma e_j(t)\Big]\Big| \\
&\leqslant 2\sum_{i=1}^{N} \Big|e_i^T(t)\hat{H}^T\hat{H}\left[f(x_i(t)) - f(\bar{x}(t))\right]\Big| \\
&\quad +2c\sum_{i=1}^{N}\sum_{j=1}^{N} |G_{ij}|\Big|e_i^T(t)\hat{H}^T\hat{H}\Gamma e_j(t)\Big|
\end{aligned}
$$

$$\leqslant c\lambda^2 \sum_{i=1}^{N}\sum_{j=1}^{N} |G_{ij}| \left(\lambda_M^2(\hat{H}^T\hat{H}) + \lambda_M(\Gamma^T\Gamma)\right)$$
$$+N\lambda^2\left(\lambda_M^2(\hat{H}^T\hat{H}) + L^2\right).$$

Thus, $\beta(t)$ is uniformly continuous.

According to Lemma 1.8, we can obtain

$$\lim_{t\to+\infty} \beta(t) = 0.$$

Then, we have

$$\lim_{t\to+\infty} \|\hat{y}_i(t)\| = \lim_{t\to+\infty} \|\hat{H}e_i(t)\| = 0, \quad i = 1, 2, \cdots, N.$$

Therefore, the complex network (3.1) achieves output synchronization. The proof is completed.

From Theorem 3.11, we can easily obtain the following conclusion. In order to obtain our result, we introduce the following assumption:

(A2) (see [1]) There exist a positive definite diagonal matrix $\hat{M} = \mathrm{diag}(\hat{m}_1, \hat{m}_2, \cdots, \hat{m}_n) \in \mathbb{R}^{n\times n}$ and a diagonal matrix $\hat{\Delta} = \mathrm{diag}(\hat{\delta}_1, \hat{\delta}_2, \cdots, \hat{\delta}_n) \in \mathbb{R}^{n\times n}$ such that $f(\cdot)$ satisfies the following inequality:

$$(x-y)^T\hat{M}[f(x) - f(y) - \hat{\Delta}(x-y)] \leqslant -\hat{\eta}(x-y)^T(x-y)$$

for some $\hat{\eta} \in \mathbb{R} > 0$ and all $x, y \in \mathbb{R}^n$.

Corollary 3.12 *Let (A2) hold. If there exist a matrix $Q \in \mathbb{R}^{rN\times rN} > 0$ and a matrix $P \in \mathbb{R}^{rN\times pN}$ such that*

$$\begin{pmatrix} \mathbb{W}_4 & \mathbb{W}_5 \\ \mathbb{W}_5^T & \mathbb{H}_3 \end{pmatrix} \leqslant 0, \tag{3.19}$$

where $\mathbb{W}_4 = 2[I_N \otimes (\hat{M}\hat{\Delta} - \hat{\eta}I_n)] + c[G \otimes (\hat{M}\Gamma + \Gamma^T\hat{M})] + (I_N \otimes \hat{H}^T)Q(I_N \otimes \hat{H})$, $\mathbb{W}_5 = I_N \otimes (\hat{M}B) - \frac{(I_N\otimes\hat{H}^T)P}{2} + (I_N \otimes \hat{H}^T)Q(I_N \otimes \hat{D})$, $\mathbb{H}_3 = (I_N \otimes \hat{D}^T)Q(I_N \otimes \hat{D}) - \frac{(I_N\otimes\hat{D}^T)P + P^T(I_N\otimes\hat{D})}{2}$, then the complex network (3.1) achieves output synchronization.

Proof Define the following Lyapunov functional for system (3.8):

$$V_2(e(t)) = \sum_{i=1}^{N} e_i^T(t)\hat{M}e_i(t).$$

In the following, we calculate the derivative $\dot{V}_2(e(t))$ along the trajectory of system (3.8):

$$\begin{aligned}\dot{V}_2(e(t)) &= 2\sum_{i=1}^{N} e_i^T(t)\hat{M}\dot{e}_i(t)\\ &= 2\sum_{i=1}^{N} e_i^T(t)\hat{M}[f(x_i(t)) - f(\bar{x}(t)) + f(\bar{x}(t))\\ &\quad -\frac{1}{N}\sum_{i=1}^{N} f(x_i(t)) + c\sum_{j=1}^{N} G_{ij}\Gamma e_j(t)\\ &\quad +Bu_i(t) - \frac{1}{N}\sum_{i=1}^{N} Bu_i(t)].\end{aligned}$$

According to (A2), we can obtain

$$\begin{aligned}&\sum_{i=1}^{N} e_i^T(t)\hat{M}(f(x_i(t)) - f(\bar{x}(t)))\\ &= \sum_{i=1}^{N}(x_i(t) - \bar{x}(t))^T\hat{M}(f(x_i(t)) - f(\bar{x}(t)))\\ &\leqslant \sum_{i=1}^{N}(x_i(t) - \bar{x}(t))^T(\hat{M}\hat{\Delta} - \hat{\eta}I_n)(x_i(t) - \bar{x}(t))\\ &= \sum_{i=1}^{N} e_i^T(t)(\hat{M}\hat{\Delta} - \hat{\eta}I_n)e_i(t).\end{aligned} \tag{3.20}$$

Because $\sum_{i=1}^{N} e_i^T(t) = 0$, we have

$$\sum_{i=1}^{N} e_i^T(t)\hat{M}(f(\bar{x}(t)) - \frac{1}{N}\sum_{i=1}^{N} f(x_i(t))) = 0, \tag{3.21}$$

$$\frac{1}{N}\sum_{i=1}^{N} e_i^T(t)\hat{M}\sum_{j=1}^{N} Bu_j(t) = 0. \tag{3.22}$$

It follows from (3.20) to (3.22) that

$$\begin{aligned}
\dot{V}_2(e(t)) &\leqslant 2\sum_{i=1}^{N} e_i^T(t)(\hat{M}\hat{\Delta} - \hat{\eta} I_n)e_i(t) \\
&\quad +2\sum_{i=1}^{N} e_i^T(t)\hat{M}Bu_i(t) \\
&\quad +2c\sum_{i=1}^{N}\sum_{j=1}^{N} G_{ij}e_i^T(t)\hat{M}\Gamma e_j(t) \\
&= e^T(t)\Big\{2[I_N \otimes (\hat{M}\hat{\Delta} - \hat{\eta} I_n)] + c[G \otimes (\hat{M}\Gamma \\
&\quad +\Gamma^T\hat{M})]\Big\}e(t) + 2e^T(t)[I_N \otimes (\hat{M}B)]u(t).
\end{aligned}$$

Therefore,

$$\begin{aligned}
&\dot{V}_2(e(t)) - \hat{y}^T(t)Pu(t) + \hat{y}^T(t)Q\hat{y}(t) \\
&\leqslant e^T(t)\Big\{2[I_N \otimes (\hat{M}\hat{\Delta} - \hat{\eta} I_n)] + c[G \otimes (\hat{M}\Gamma \\
&\quad +\Gamma^T\hat{M})]\Big\}e(t) + 2e^T(t)[I_N \otimes (\hat{M}B)]u(t) \\
&\quad -[(I_N \otimes \hat{H})e(t) + (I_N \otimes \hat{D})u(t)]^T Pu(t) \\
&\quad +[(I_N \otimes \hat{H})e(t) + (I_N \otimes \hat{D})u(t)]^T Q[(I_N \otimes \hat{H})e(t) \\
&\quad +(I_N \otimes \hat{D})u(t)] \\
&= \begin{pmatrix} e(t) \\ u(t) \end{pmatrix}^T \begin{pmatrix} \mathbb{W}_4 & \mathbb{W}_5 \\ \mathbb{W}_5^T & \mathbb{H}_3 \end{pmatrix} \begin{pmatrix} e(t) \\ u(t) \end{pmatrix}.
\end{aligned} \tag{3.23}$$

From (3.19) and (3.23), we have

$$\dot{V}_2(e(t)) \leqslant \hat{y}^T(t)Pu(t) - \hat{y}^T(t)Q\hat{y}(t). \tag{3.24}$$

By integrating equation (3.24) with respect to t over the time period t_0 to t_p, we can obtain

$$\int_{t_0}^{t_p} (\hat{y}^T(t)Pu(t) - \hat{y}^T(t)Q\hat{y}(t))dt \geqslant V_2(e(t_p)) - V_2(e(t_0))$$

for any $t_p, t_0 \in \mathbb{R}^+$ and $t_p \geqslant t_0$.

Obviously, system (3.8) is output-strictly passive in the sense of Definition 1.3.

On the other hand, according to the definition of $V_2(e(t))$, we can get

$$\lambda_m(\hat{M})\|e(t)\|^2 \leqslant V_2(e(t)) \leqslant \lambda_M(\hat{M})\|e(t)\|^2.$$

Therefore, we can conclude from Theorem 3.11 that the complex network (3.1) achieves output synchronization. The proof is completed.

Remark 3.13 In recent years, some researchers have studied the output synchronization problem of CDNs by using the passivity property [33, 35, 93–95]. But, in these papers, they always assume that the input vector has the same dimension as the output vector. However, in many practical situations, the input and output vectors have different dimensions. To the best of our knowledge, this is the first time to consider the relationship between output strict passivity and output synchronization of the CDNs with different dimensions of input and output.

3.4 Passivity and Output Synchronization of CDNs with Adaptive Coupling Strength

In recent years, the passivity of CDNs with fixed coupling strength has been extensively studied by the researchers. To our knowledge, in many real-world networks, the coupling strength is adaptively adjusted according to changes in the environment or the network itself (for instance, neural networks, wireless sensor networks, biological networks). However, very few researchers have investigated the passivity of CDNs with adaptive coupling strength. In this section, an adaptive strategy to tune the coupling strength is designed. By utilizing the designed adaptive law, several sufficient conditions ensuring passivity, input strict passivity and output strict passivity are obtained. In addition, we also discuss the relationship between output strict passivity and output synchronization of the complex dynamical network with adaptive coupling strength.

3.4.1 *Passivity Analysis of Complex Dynamical Network*

Theorem 3.14 *Let (A1) hold. If there exist a positive constant* $\hat{c}$ *and a matrix* $P \in \mathbb{R}^{qN \times pN}$ *such that*

$$\begin{pmatrix} \mathbb{W}_6 & \mathbb{H}_4 \\ \mathbb{H}_4^T & -\frac{(I_N \otimes D^T)P + P^T(I_N \otimes D)}{2} \end{pmatrix} \leqslant 0, \tag{3.25}$$

where $\mathbb{W}_6 = 2[I_N \otimes (M\Delta - \eta I_n)] + \hat{c}[G \otimes (M\Gamma + \Gamma^T M)]$, $\mathbb{H}_4 = I_N \otimes (MB) - \frac{(I_N \otimes C^T)P}{2}$, *then the complex network* (3.1) *is passive under the following adaptive law:*

$$\dot{c}(t) = \beta \sum_{i=1}^{N} \sum_{\substack{j=1 \\ j \neq i}}^{N} G_{ij}(x_i(t) - x_j(t))^T (M\Gamma + \Gamma^T M)(x_i(t) - x_j(t)), \quad (3.26)$$

where β is a positive real number.

Proof The complex network (3.1) with adaptive coupling strength can be described by

$$\begin{cases} \dot{x}_i(t) = f(x_i(t)) + c(t) \sum_{j=1}^{N} G_{ij} \Gamma x_j(t) + Bu_i(t), \\ y_i(t) = \quad Cx_i(t) + Du_i(t), \quad t \geqslant 0, \end{cases} \quad (3.27)$$

where $i = 1, 2, \cdots, N$.

Define the following Lyapunov functional for the system (3.27):

$$V_3(t) = \sum_{i=1}^{N} x_i^T(t) M x_i(t) + \frac{1}{4\beta}(c(t) - \hat{c})^2.$$

In the following, we calculate the derivative $\dot{V}_3(t)$ along the trajectory of system (3.27):

$$\begin{aligned} \dot{V}_3(t) &= 2 \sum_{i=1}^{N} x_i^T(t) M \dot{x}_i(t) + \frac{1}{2\beta}(c(t) - \hat{c})\dot{c}(t) \\ &= 2 \sum_{i=1}^{N} x_i^T(t) M [f(x_i(t)) + c(t) \sum_{j=1}^{N} G_{ij} \Gamma x_j(t) \\ &\quad + Bu_i(t)] + \frac{c(t) - \hat{c}}{2} \sum_{i=1}^{N} \sum_{\substack{j=1 \\ j \neq i}}^{N} G_{ij}(x_i(t) \\ &\quad - x_j(t))^T (M\Gamma + \Gamma^T M)(x_i(t) - x_j(t)) \\ &\leqslant 2 \sum_{i=1}^{N} x_i^T(t)(M\Delta - \eta I_n) x_i(t) \\ &\quad + 2 \sum_{i=1}^{N} x_i^T(t) M B u_i(t) \\ &\quad + 2c(t) \sum_{i=1}^{N} \sum_{j=1}^{N} G_{ij} x_i^T(t) M \Gamma x_j(t) \end{aligned}$$

$$-(c(t)-\hat{c})\sum_{i=1}^{N}\sum_{j=1}^{N}G_{ij}x_i^T(t)(M\Gamma+\Gamma^T M)x_j(t)$$

$$=2\sum_{i=1}^{N}x_i^T(t)(M\Delta-\eta I_n)x_i(t)$$

$$+\hat{c}\sum_{i=1}^{N}\sum_{j=1}^{N}G_{ij}x_i^T(t)(M\Gamma+\Gamma^T M)x_j(t)$$

$$+2\sum_{i=1}^{N}x_i^T(t)MBu_i(t).$$

Let $x(t)=(x_1^T(t),\ x_2^T(t),\ \cdots,\ x_N^T(t))^T$, $u(t)=(u_1^T(t),u_2^T(t),\cdots,u_N^T(t))^T$ and $y(t)=(y_1^T(t),y_2^T(t),\cdots,y_N^T(t))^T$. Then, we have

$$\dot{V}_3(t)-y^T(t)Pu(t)$$

$$\leqslant 2\sum_{i=1}^{N}x_i^T(t)(M\Delta-\eta I_n)x_i(t)$$

$$+\hat{c}\sum_{i=1}^{N}\sum_{j=1}^{N}G_{ij}x_i^T(t)(M\Gamma+\Gamma^T M)x_j(t)$$

$$+2\sum_{i=1}^{N}x_i^T(t)MBu_i(t)$$

$$-[(I_N\otimes C)x(t)+(I_N\otimes D)u(t)]^T Pu(t)$$

$$=\xi^T(t)\begin{pmatrix}\mathbb{W}_6 & \mathbb{H}_4\\ \mathbb{H}_4^T & -\frac{(I_N\otimes D^T)P+P^T(I_N\otimes D)}{2}\end{pmatrix}\xi(t),$$

where $\xi(t)=(x^T(t),u^T(t))^T$.

From (3.25), we have

$$\dot{V}_3(t)\leqslant y^T(t)Pu(t). \tag{3.28}$$

By integrating equation (3.28) with respect to t over the time period t_0 to t_p, we can obtain

$$\int_{t_0}^{t_p}y^T(t)Pu(t)dt\geqslant V_3(t_p)-V_3(t_0)$$

for any $t_p, t_0 \in \mathbb{R}^+$ and $t_p \geqslant t_0$. The proof is completed.

By the similar proof of Theorem 3.14, we can obtain the following conclusions. Here we omit their proof to avoid the repetition.

Theorem 3.15 *Let (A1) hold. If there exist a matrix $W \in \mathbb{R}^{pN\times pN} > 0$, a matrix $P \in \mathbb{R}^{qN\times pN}$, and a positive constant $\hat{c}$ such that*

$$\begin{pmatrix} \mathbb{W}_6 & \mathbb{H}_4 \\ \mathbb{H}_4^T & W - \frac{(I_N\otimes D^T)P+P^T(I_N\otimes D)}{2} \end{pmatrix} \leqslant 0, \tag{3.29}$$

where $\mathbb{W}_6 = 2[I_N \otimes (M\Delta - \eta I_n)] + \hat{c}[G \otimes (M\Gamma + \Gamma^T M)]$, $\mathbb{H}_4 = I_N \otimes (MB) - \frac{(I_N\otimes C^T)P}{2}$, then the complex network (3.1) *is input-strictly passive under the following adaptive law:*

$$\dot{c}(t) = \beta \sum_{i=1}^{N}\sum_{\substack{j=1\\ j\neq i}}^{N} G_{ij}(x_i(t) - x_j(t))^T(M\Gamma + \Gamma^T M)(x_i(t) - x_j(t)),$$

where β is a positive real number.

Theorem 3.16 *Let (A1) hold. If there exist a matrix $Q \in \mathbb{R}^{qN\times qN} > 0$, a matrix $P \in \mathbb{R}^{qN\times pN}$, and a positive constant $\hat{c}$ such that*

$$\begin{pmatrix} \mathbb{W}_7 & \mathbb{W}_8 \\ \mathbb{W}_8^T & \mathbb{H}_5 \end{pmatrix} \leqslant 0, \tag{3.30}$$

where $\mathbb{W}_7 = 2[I_N \otimes (M\Delta - \eta I_n)] + \hat{c}[G \otimes (M\Gamma + \Gamma^T M)] + (I_N \otimes C^T)Q(I_N \otimes C)$, $\mathbb{W}_8 = I_N \otimes (MB) - \frac{(I_N\otimes C^T)P}{2} + (I_N \otimes C^T)Q(I_N \otimes D)$, $\mathbb{H}_5 = (I_N \otimes D^T)Q(I_N \otimes D) - \frac{(I_N\otimes D^T)P+P^T(I_N\otimes D)}{2}$, then the complex network (3.1) *is output-strictly passive under the following adaptive law:*

$$\dot{c}(t) = \beta \sum_{i=1}^{N}\sum_{\substack{j=1\\ j\neq i}}^{N} G_{ij}(x_i(t) - x_j(t))^T(M\Gamma + \Gamma^T M)(x_i(t) - x_j(t)),$$

where β is a positive real number.

Remark 3.17 In [97], Li, Shi and Yao designed some effective adaptive laws for adjusting the parameters of the sliding mode controller. To our knowledge, very few researchers have discussed the passivity of CDNs with adaptive coupling strength. Especially, the passivity of complex networks with adaptive coupling strength and different dimensions of input and output has not yet been investigated. In Theorems 3.14, 3.15, and 3.16, several criteria for passivity, input strict passivity and output strict passivity are derived by using the designed adaptive law.

3.4.2 Passivity-Based Output Synchronization of Complex Dynamical Network

In this subsection, we reveal the relationship between output strict passivity and output synchronization of the complex dynamical network (3.1) with adaptive coupling strength. In addition, by utilizing the relationship between output strict passivity and output synchronization, a criterion for output synchronization is established.

Here, consider the case that the coupling strength c is updated according to the following adaptive law:

$$\dot{c}(t) = \beta \sum_{i=1}^{N} \sum_{\substack{j=1 \\ j\neq i}}^{N} G_{ij}(x_i(t) - x_j(t))^T(\tilde{M}\Gamma + \Gamma^T \tilde{M})(x_i(t) \\ - x_j(t)), \tag{3.31}$$

where β is a positive real number, $\tilde{M} \in \mathbb{R}^{n\times n}$. Then, we can get from (3.8) that

$$\begin{cases} \dot{e}_i(t) = & f(x_i(t)) - \frac{1}{N}\sum_{i=1}^{N} f(x_i(t)) + Bu_i(t) \\ & +c(t)\sum_{j=1}^{N} G_{ij}\Gamma e_j(t) - \sum_{i=1}^{N} \frac{Bu_i(t)}{N}, \\ \dot{c}(t) = & \beta \sum_{i=1}^{N} \sum_{\substack{j=1 \\ j\neq i}}^{N} G_{ij}(e_i(t) - e_j(t))^T(\tilde{M}\Gamma \\ & +\Gamma^T\tilde{M})(e_i(t) - e_j(t)), \\ \hat{y}_i(t) = & \hat{H}e_i(t) + \hat{D}u_i(t), \end{cases} \tag{3.32}$$

where $i = 1, 2, \cdots, N$.

For the convenience, we denote

$$\begin{aligned} u(t) &= (u_1^T(t), u_2^T(t), \cdots, u_N^T(t))^T, \\ \hat{y}(t) &= (\hat{y}_1^T(t), \hat{y}_2^T(t), \cdots, \hat{y}_N^T(t))^T, \\ e(t) &= (e_1^T(t), e_2^T(t), \cdots, e_N^T(t))^T, \\ \hat{e}(t) &= (e^T(t), c(t))^T. \end{aligned}$$

Theorem 3.18 *The complex network* (3.1) *achieves output synchronization under the adaptive law* (3.31) *if matrix* $\tilde{M}\Gamma + \Gamma^T\tilde{M} > 0$ *and system* (3.32) *is output-strictly passive with respect to storage function*

$$V_4(t) = \sum_{i=1}^{N} e_i^T(t)Ze_i(t) + \frac{1}{4\beta}(c(t) - \hat{c})^2,$$

where $Z \in \mathbb{R}^{n\times n} > 0$, $\hat{c} \in \mathbb{R} > 0$.

Proof If system (3.32) is output-strictly passive with respect to storage function $V_4(t)$, then there exist a matrix $Q \in \mathbb{R}^{rN \times rN} > 0$ and a matrix $P \in \mathbb{R}^{rN \times pN}$ such that

$$\int_t^{t+\varepsilon} (\hat{y}^T(s)Pu(s) - \hat{y}^T(s)Q\hat{y}(s))ds \geqslant V_4(t+\varepsilon) - V_4(t) \tag{3.33}$$

for any $t \in \mathbb{R}^+$ and $\varepsilon > 0$.

From (3.33), we can obtain

$$\begin{aligned} \frac{V_4(t+\varepsilon) - V_4(t)}{\varepsilon} \leqslant & \frac{\int_t^{t+\varepsilon} \hat{y}^T(s)Pu(s)ds}{\varepsilon} \\ & - \frac{\int_t^{t+\varepsilon} \hat{y}^T(s)Q\hat{y}(s)ds}{\varepsilon}. \end{aligned} \tag{3.34}$$

By taking limit $\epsilon \to 0$ in (3.34), we have

$$\dot{V}_4(t) \leqslant \hat{y}^T(t)Pu(t) - \hat{y}^T(t)Q\hat{y}(t). \tag{3.35}$$

Letting $u_i(t) = 0, i = 1, 2, \cdots, N$, we can get from (3.35) that

$$\begin{aligned} \dot{V}_4(t) &\leqslant -e^T(t)(I_N \otimes \hat{H}^T)Q(I_N \otimes \hat{H})e(t) \\ &\leqslant -\lambda_m(Q) \sum_{i=1}^{N} e_i^T(t)\hat{H}^T\hat{H}e_i(t). \end{aligned} \tag{3.36}$$

Thus, it follows that $V_4(t)$ is nonincreasing, and so each term of $V_4(t)$ is bounded. Consequently, both the error vector $e(t)$ and the coupling strength $c(t)$ are bounded. Since $c(t)$ is monotonically increasing (see (3.31)), we have

$$\lim_{t \to +\infty} c(t) = c_0 \in \mathbb{R} > 0.$$

In what follows, by the similar proof of Theorem 3.11, we can easily obtain

$$\lim_{t \to +\infty} \|\hat{y}_i(t)\| = 0, \quad i = 1, 2, \cdots, N.$$

Therefore, the complex network (3.1) achieves output synchronization. The proof is completed.

From Theorem 3.18, we can easily obtain the following conclusion.

Corollary 3.19 *Let (A2) hold and $\hat{M}\Gamma + \Gamma^T\hat{M} > 0$. If there exist a matrix $Q \in \mathbb{R}^{rN \times rN} > 0$, a matrix $P \in \mathbb{R}^{rN \times pN}$ and a positive constant $\hat{c}$ such that*

$$\begin{pmatrix} \mathbb{W}_9 & \mathbb{W}_{10} \\ \mathbb{W}_{10}^T & \mathbb{H}_6 \end{pmatrix} \leqslant 0, \tag{3.37}$$

where $\mathbb{W}_{10} = I_N \otimes (\hat{M}B) - \frac{(I_N \otimes \hat{H}^T)P}{2} + (I_N \otimes \hat{H}^T)Q(I_N \otimes \hat{D})$, $\mathbb{W}_9 = 2[I_N \otimes (\hat{M}\hat{\Delta} - \hat{\eta}I_n)] + \hat{c}[G \otimes (\hat{M}\Gamma + \Gamma^T \hat{M})] + (I_N \otimes \hat{H}^T)Q(I_N \otimes \hat{H})$, $\mathbb{H}_6 = (I_N \otimes \hat{D}^T)Q(I_N \otimes \hat{D}) - \frac{(I_N \otimes \hat{D}^T)P + P^T(I_N \otimes \hat{D})}{2}$, *then the complex network* (3.1) *achieves output synchronization under the following adaptive law:*

$$\begin{aligned} \dot{c}(t) = \beta \sum_{i=1}^{N} \sum_{\substack{j=1 \\ j \neq i}}^{N} G_{ij}(x_i(t) - x_j(t))^T (\hat{M}\Gamma + \Gamma^T \hat{M})(x_i(t) \\ -x_j(t)), \end{aligned} \tag{3.38}$$

where β *is a positive real number.*

Proof By (3.32) and (3.38), we can get

$$\begin{cases} \dot{e}_i(t) = & f(x_i(t)) - \frac{1}{N}\sum_{i=1}^{N} f(x_i(t)) + Bu_i(t) \\ & +c(t)\sum_{j=1}^{N} G_{ij}\Gamma e_j(t) - \sum_{i=1}^{N} \frac{Bu_i(t)}{N}, \\ \dot{c}(t) = & \beta \sum_{i=1}^{N} \sum_{\substack{j=1 \\ j\neq i}}^{N} G_{ij}(e_i(t) - e_j(t))^T (\hat{M}\Gamma \\ & +\Gamma^T \hat{M})(e_i(t) - e_j(t)), \\ \hat{y}_i(t) = & \hat{H}e_i(t) + \hat{D}u_i(t), \end{cases} \tag{3.39}$$

where $i = 1, 2, \cdots, N$.

Define the following Lyapunov functional for system (3.39):

$$V_5(t) = \sum_{i=1}^{N} e_i^T(t)\hat{M}e_i(t) + \frac{1}{4\beta}(c(t) - \hat{c})^2.$$

In the following, we calculate the derivative $\dot{V}_5(t)$ along the trajectory of system (3.39):

$$\begin{aligned} \dot{V}_5(t) =& 2\sum_{i=1}^{N} e_i^T(t)\hat{M}\dot{e}_i(t) + \frac{c(t) - \hat{c}}{2} \sum_{i=1}^{N} \sum_{\substack{j=1 \\ j\neq i}}^{N} G_{ij}(e_i(t) \\ & -e_j(t))^T (\hat{M}\Gamma + \Gamma^T \hat{M})(e_i(t) - e_j(t)) \\ =& 2\sum_{i=1}^{N} e_i^T(t)\hat{M}[f(x_i(t)) - f(\bar{x}(t)) + f(\bar{x}(t)) \end{aligned}$$

$$
\begin{aligned}
& -\frac{1}{N}\sum_{i=1}^{N} f(x_i(t)) + c(t)\sum_{j=1}^{N} G_{ij}\Gamma e_j(t) + Bu_i(t) \\
& -\frac{1}{N}\sum_{i=1}^{N} Bu_i(t)] \\
& -(c(t)-\hat{c})\sum_{i=1}^{N}\sum_{j=1}^{N} G_{ij}e_i^T(t)(\hat{M}\Gamma + \Gamma^T\hat{M})e_j(t) \\
\leqslant & 2\sum_{i=1}^{N} e_i^T(t)(\hat{M}\hat{\Delta} - \hat{\eta}I_n)e_i(t) \\
& +\hat{c}\sum_{i=1}^{N}\sum_{j=1}^{N} G_{ij}e_i^T(t)(\hat{M}\Gamma + \Gamma^T\hat{M})e_j(t) \\
& +2\sum_{i=1}^{N} e_i^T(t)\hat{M}Bu_i(t).
\end{aligned}
$$

Therefore,

$$
\begin{aligned}
& \dot{V}_5(t) - \hat{y}^T(t)Pu(t) + \hat{y}^T(t)Q\hat{y}(t) \\
\leqslant & e^T(t)\Big\{2[I_N \otimes (\hat{M}\hat{\Delta} - \hat{\eta}I_n)] + \hat{c}[G \otimes (\hat{M}\Gamma \\
& +\Gamma^T\hat{M})]\Big\}e(t) + 2e^T(t)[I_N \otimes (\hat{M}B)]u(t) \\
& -[(I_N \otimes \hat{H})e(t) + (I_N \otimes \hat{D})u(t)]^T Pu(t) \\
& +[(I_N \otimes \hat{H})e(t) + (I_N \otimes \hat{D})u(t)]^T Q[(I_N \otimes \hat{H})e(t) \\
& +(I_N \otimes \hat{D})u(t)] \\
= & \begin{pmatrix} e(t) \\ u(t) \end{pmatrix}^T \begin{pmatrix} \mathbb{W}_9 & \mathbb{W}_{10} \\ \mathbb{W}_{10}^T & \mathbb{H}_6 \end{pmatrix} \begin{pmatrix} e(t) \\ u(t) \end{pmatrix}.
\end{aligned}
$$

From (3.37), we have

$$
\dot{V}_5(t) \leqslant \hat{y}^T(t)Pu(t) - \hat{y}^T(t)Q\hat{y}(t). \tag{3.40}
$$

By integrating equation (3.40) with respect to t over the time period t_0 to t_p, we can obtain

$$
\int_{t_0}^{t_p} (\hat{y}^T(t)Pu(t) - \hat{y}^T(t)Q\hat{y}(t))dt \geqslant V_5(t_p) - V_5(t_0)
$$

for any $t_p, t_0 \in \mathbb{R}^+$ and $t_p \geqslant t_0$.

Therefore, system (3.39) is output-strictly passive with respect to storage function $V_5(t)$.

According to Theorem 3.18, the complex network (3.1) achieves output synchronization under the adaptive law (3.38). The proof is completed.

Remark 3.20 In recent years, some authors have investigated the output synchronization problem for CDNs using the passivity property of each node [98, 99]. In these existing works, it is assumed that the input vector has the same dimension as the output vector. Moreover, the interconnection between nodes is realized by the designed control input [100–102]. The main difficulty of this chapter is to analyze the relationship between output strict passivity and output synchronization of CDNs with different dimensions of input and output vectors, which can not be dealt with by those methods used in traditional CDNs. By utilizing Barbalat's lemma, some sufficient conditions for output synchronization of the output-strictly passive CDNs with fixed and adaptive coupling strength are established, respectively.

Remark 3.21 In this section, some adaptive strategies to tune the coupling strength c are designed to guarantee the passivity and output synchronization of the CDNs. By utilizing the adaptive strategies defined by the error vectors, the coupling strength c can be quickly adjusted. Moreover, the coupling strength c can be adjusted to a suitable value in Theorem 3.18 and Corollary 3.19

3.5 Numerical Examples

In this section, we give two examples and their simulation to show the effectiveness of the above obtained theoretical results.

Example 3.22 Consider a complex dynamical network consisting of five identical nodes, in which each node is a 3-dimensional linear system described by [8]

$$\begin{pmatrix} \dot{x}_{i1} \\ \dot{x}_{i2} \\ \dot{x}_{i3} \end{pmatrix} = \begin{pmatrix} -x_{i1} \\ -2x_{i2} \\ -3x_{i3} \end{pmatrix}$$

where $i = 1, 2, \cdots, 5$. It is easy to verify that (A2) is satisfied with $\hat{M} = I_3$, $\hat{\Delta} = 0$ and $\hat{\eta} = 1$.

We take $c = 0.4$. The matrices B, C, D, Γ and G are chosen as, respectively,

$$B = \begin{pmatrix} 0.2 \\ 0.4 \\ 0.1 \end{pmatrix},$$

$$\Gamma = \begin{pmatrix} 0.6 & 0.8 & 0.7 \\ 0.9 & 0.6 & 0.5 \\ 0.8 & 0.5 & 0.7 \end{pmatrix},$$

$$C = \begin{pmatrix} 1 & 0 & 0 \\ 0 & 1 & 0 \end{pmatrix},$$

$$D = \begin{pmatrix} 0.4 \\ 0.2 \end{pmatrix},$$

$$G = \begin{pmatrix} -1.3 & 0.6 & 0.7 & 0 & 0 \\ 0.6 & -1.4 & 0.8 & 0 & 0 \\ 0.7 & 0.8 & -2.9 & 0.5 & 0.9 \\ 0 & 0 & 0.5 & -0.5 & 0 \\ 0 & 0 & 0.9 & 0 & -0.9 \end{pmatrix}.$$

By using the YALMIP Toolbox of MATLAB, we can find the following matrices

$$P = I_5 \otimes \begin{pmatrix} 1.9310 \\ 1.8906 \end{pmatrix},$$

$$Q = I_5 \otimes \begin{pmatrix} 0.7467 & 0.4620 \\ 0.4620 & 0.7811 \end{pmatrix} > 0$$

satisfying (3.19). According to Corollary 3.12, the complex network (3.1) with above given parameters achieves output synchronization. The evolutions of the output variables of five nodes are shown in Fig. 3.1.

Example 3.23 Consider a complex dynamical network consisting of five identical nodes, in which each node is a 3-dimensional linear system described by

$$\begin{pmatrix} \dot{x}_{i1} \\ \dot{x}_{i2} \\ \dot{x}_{i3} \end{pmatrix} = \begin{pmatrix} -x_{i1} \\ -2x_{i2} \\ -3x_{i3} \end{pmatrix}$$

where $i = 1, 2, \cdots, 5$. It is easy to verify that (A2) is satisfied with $\hat{M} = I_3$, $\hat{\Delta} = 0$ and $\hat{\eta} = 1$.

We take $c(0) = 0.001$. The matrices B, C, D, Γ and G are chosen as, respectively,

$$B = \begin{pmatrix} 0.3 \\ 0.2 \\ 0.4 \end{pmatrix},$$

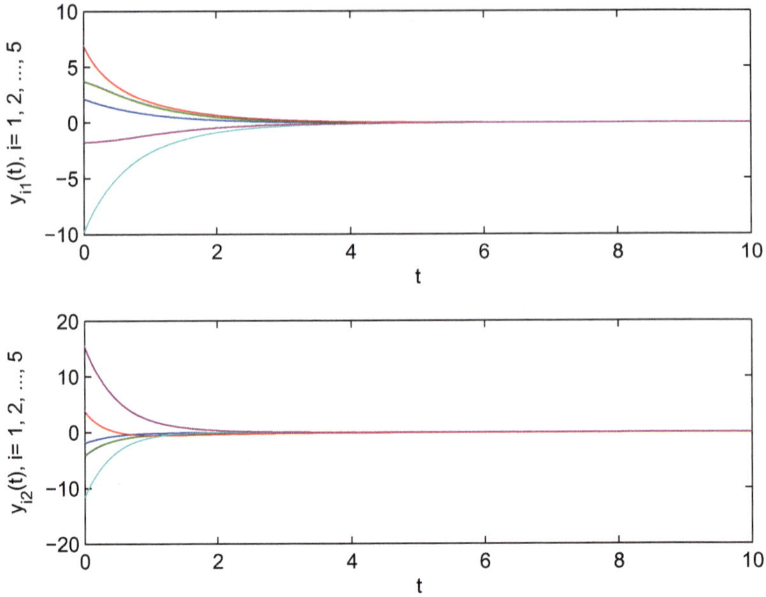

Fig. 3.1 The change processes of the output variables of complex dynamical network (3.1) in time interval [0, 10]

$$\Gamma = \begin{pmatrix} 0.9 & 0 & 0 \\ 0 & 0.8 & 0 \\ 0 & 0 & 1.2 \end{pmatrix},$$

$$C = \begin{pmatrix} 1 & 0 & 0 \\ 0 & 1 & 0 \end{pmatrix},$$

$$D = \begin{pmatrix} 0.3 \\ 0.5 \end{pmatrix},$$

$$G = \begin{pmatrix} -1.5 & 0.7 & 0.8 & 0 & 0 \\ 0.7 & -1.2 & 0.5 & 0 & 0 \\ 0.8 & 0.5 & -2.4 & 0.6 & 0.5 \\ 0 & 0 & 0.6 & -0.6 & 0 \\ 0 & 0 & 0.5 & 0 & -0.5 \end{pmatrix}.$$

Obviously, $\hat{M}\Gamma + \Gamma^T\hat{M} > 0$.
By using the YALMIP Toolbox of MATLAB, we can find the following matrices

$$P = I_5 \otimes \begin{pmatrix} 1.4470 \\ 1.8055 \end{pmatrix},$$

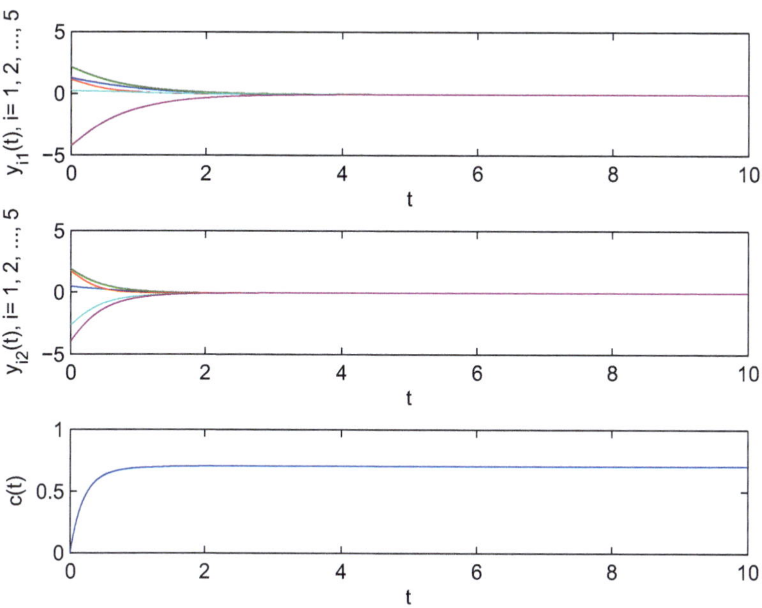

Fig. 3.2 The change processes of $y_{i1}(t)$, $y_{i2}(t)(i = 1, 2, \cdots, 5)$ and $c(t)$

$$Q = I_5 \otimes \begin{pmatrix} 0.5993 & 0.0840 \\ 0.0840 & 0.6828 \end{pmatrix} > 0$$

satisfying (3.37) with $\hat{c} = 0.0296$. According to Corollary 3.19, the complex network (3.1) with above given parameters achieves output synchronization under the adaptive law (3.38). The simulation results are shown in Fig. 3.2.

From Figs. 3.1 and 3.2, we clearly see that the output variables $y_{i1}(t)$ and $y_{i2}(t)$ $(i = 1, 2, \cdots, 5)$ are very close to 0 when the time t increases gradually to 2s, and this state is maintained along with the increasing of the time. These results show that the complex network achieves output synchronization. On the other hand, Fig. 3.2 clearly show that the coupling strength $c(t)$ asymptotically converges to a finite positive value.

3.6 Conclusion

In this chapter, a complex dynamical network with different dimensions of input and output has been introduced. The passivity, input strict passivity and output strict passivity of the proposed network model with fixed and adaptive coupling strength have been taken into consideration, and several sufficient conditions have been established. Furthermore, the relationship between output strict passivity and output synchronization of the proposed network model with fixed and adaptive coupling strength have been revealed. In addition, some sufficient conditions ensuring output synchronization have been derived by utilizing the relationship between output strict passivity and output synchronization. Two numerical examples have been provided to verify the correctness and effectiveness of the theoretical results. In future work, we shall consider the cases that the coupling weights G_{ij} are adaptively adjusted and the topological structure of complex network is directed.

Chapter 4
Analysis and Control of Output Synchronization in Directed and Undirected CDNs

4.1 Introduction

In recent years, analyses of the dynamical behavior and control problems of CDNs have been extensively carried out due to the fact that many natural and artificial systems, such as citation networks, metabolic networks, Internet, biological neural networks, CNNs and food webs, can be described as complex network models. The investigation of CDNs obviously plays a prominent role both in application and theory, and a considerable amount of significant results have been reported about the dynamical behavior of complex networks.

In particular, the synchronization in complex networks has drawn much attention, and several synchronization criteria have been established for different kinds of CDNs [8, 9, 17, 18, 22, 23, 56, 60, 61, 88–90, 92, 103, 104]. For instance, Lü and Chen [8] introduced a time-varying complex dynamical network and studied the synchronization of the proposed network model. In [9], the authors discussed the pinning synchronization of an undirected network model, and a criterion for the synchronization of such network was derived. In addition, some adaptive control schemes for updating the feedback gains and coupling strength were presented. Zhao et al. [90] established a few conditions for the synchronization of switched CDNs. Jiang et al. [22] put forward a complex dynamical network model, in which the nodes are coupled through their output vectors, and then established some synchronization conditions based on Lyapunov stability theory. In [23], the authors studied the synchronization of a complex dynamical network, in which they considered not only output coupling but also random sensor delay. In these existing results [8, 9, 17, 18, 22, 23, 56, 60, 61, 90, 103, 104], the authors considered only the influence of time on the node state. In many applications, however, it is also of great significance to consider the space variable [88, 89, 92]. Yang et al. [92] investigated the exponential synchronization problem for a delayed CNNs with reaction-diffusion terms by applying a pinning-impulsive control scheme. In [89], Wang and Wu presented a CNNs with reaction-diffusion terms and hybrid coupling

J.-L. Wang et al., *Analysis and Control of Output Synchronization for Complex Dynamical Networks*, https://doi.org/10.1007/978-981-13-1352-3_4

and considered the $\mathcal{H}_\infty$ synchronization and synchronization, respectively, of the proposed network model.

It is important to note that the state synchronization in CDNs has been previously investigated [8, 9, 17, 18, 22, 23, 56, 60, 61, 88–90, 92, 103, 104]. State synchronization is a highly preferable property but a very difficult problem. Furthermore, in many instances, the synchronization is based only on a small section of states. Thus, it is more meaningful to consider the output synchronization in CDNs. Recently, a number of researchers have focused on multi-agent systems and investigated the problem of output synchronization [33, 35, 98, 105, 106]. In [33, 35, 98, 105], the authors focused on multi-agent systems and considered the output synchronization by using the passivity property of each agent. Grip et al. [106] discussed the output synchronization in heterogeneous networks of linear agents. However, only a few researchers have studied the problem of output synchronization in CDNs [24, 30]. In [24], a few conditions for global and local exponential output synchronization in a CDN with output coupling were presented. By applying the designed adaptive output feedback controllers, some sufficient conditions were established to ensure the output synchronization in CDNs [30].

The coupling weights are well known to be a key factor that has an impact on the state synchronization in CDNs. The appropriate coupling weights can apparently enhance the synchronizability of the networks [107]. Further, in many practical networks (such as wireless sensor networks, biological networks, and neural networks), the coupling weights among network nodes are adjusted adaptively based on changes in the surrounding circumstances or the network itself [85, 108]. Hence, it is of great significance to investigate the output synchronization problem in CDNs with adaptive coupling weights. In the last few years, several adaptive strategies for updating the coupling weights in network models have been developed to achieve synchronization [11, 55, 88, 89, 107–111]. Zhou and Kurths [107] introduced an adaptive scheme based on a local synchronization property, which contributed to the global state synchronization in the whole network. In [108], node and edge-based strategies, as two local adaptive strategies, were designed for the state synchronization of CDNs. Two classes of edge-based adaptive laws for ensuring network synchronization were introduced in [11, 109]. In [110], the authors presented a decentralized adaptive strategy to discuss the synchronization and pinning control of a network of Lur'e systems. Wang et al. [88] investigated the state synchronization of two CNNs with reaction-diffusion terms and an undirected topology by using an edge-based adaptive strategy. However, these existing works [11, 55, 88, 89, 107–111] mainly discussed the state synchronization of CDNs. To the best of our knowledge, only a few researchers have focused on the output synchronization of CDNs with adaptive coupling weights.

The present research investigates the output synchronization of CDNs with adaptive and fixed coupling weights, respectively, and contributes to the literature in a number of ways. First, based on Barbalat's lemma, a sufficient condition is constructed to ensure the output synchronization of a complex dynamical network with a directed topology and fixed coupling weights. Second, an output synchronization criterion is established for a complex dynamical network with an undirected topology and fixed coupling weights, which extends the existing results. Third, an analysis of the output synchronization of a complex dynamical network with a directed topology and adaptive coupling weights is carried out, and a design procedure of the adaptive law is presented. Fourth, an adaptive strategy is designed to tune the coupling weights of a complex dynamical network with an undirected topology, and the outputs of a complex dynamical network are proven to achieve synchronization on the condition that the coupling weights are adjusted according to the designed adaptive law.

4.2 Network Model

The CDN model is given as follows:

$$\begin{cases} \dot{x}_i(t) = f(x_i(t)) + c\sum_{j=1}^{N} G_{ij}\Gamma x_j(t), \\ y_i(t) = \quad Cx_i(t), \quad t \geqslant 0, \end{cases} \tag{4.1}$$

where $i = 1, 2, \cdots, N$; $f(\cdot) \in \mathbb{R}^n$ denotes a continuously differentiable vector function; $x_i(t) = (x_{i1}(t), x_{i2}(t), \cdots, x_{in}(t))^T \in \mathbb{R}^n$ represents the state vector of node i; $y_i(t) = (y_{i1}(t), y_{i2}(t), \cdots, y_{iq}(t))^T \in \mathbb{R}^q$ refers to the output vector of node i; C is a known matrix with proper dimensions; $0 < \Gamma = \mathrm{diag}(\gamma_1, \gamma_2, \cdots, \gamma_n) \in \mathbb{R}^{n\times n}$ is the inner coupling matrix; and c is a positive real number representing the overall coupling strength. $G = (G_{ij})_{N\times N}$ is the coupling configuration matrix representing the topological structure: if there is a connection from nodes i to j, then $G_{ij} > 0$; otherwise, $G_{ij} = 0 (i \neq j)$. In addition, G_{ii} satisfies

$$G_{ii} = -\sum_{\substack{j=1 \\ j\neq i}}^{N} G_{ij}.$$

In this chapter, the function $f(\cdot)$ is in the QUAD function class [43, 112]; that is, there exist diagonal matrices $0 < P = \mathrm{diag}(p_1, p_2, \cdots, p_n) \in \mathbb{R}^{n\times n}$ and $\Delta = \mathrm{diag}(\delta_1, \delta_2, \cdots, \delta_n) \in \mathbb{R}^{n\times n}$, such that

$$\begin{aligned} &(x-y)^T P[f(x) - f(y) - \Delta(x-y)] \qquad \textbf{(A1)} \\ &\leqslant -\eta(x-y)^T(x-y) \end{aligned}$$

for all $x, y \in \mathbb{R}^n$ and some $0 < \eta \in \mathbb{R}$.

Remark 4.1 In this chapter, complex dynamical network (4.1) is always assumed to be connected. However, the coupling configurations are not required to be the symmetric connections. In addition, the output matrix C is an arbitrary $q \times n$ real matrix. To ensure the uniqueness and existence of the solutions for network (4.1), the function $f(\cdot)$ is assumed to satisfy the Lipschitz condition. This assumption has been extensively used to analyze the synchronization of CDNs [104], which not only can fully ensure the uniqueness and existence of the solutions for complex networks but also facilitates the analysis of the dynamical behaviors of such networks.

Definition 4.2 (see [24]) The outputs of the network model (4.1) are considered to achieve synchronization if they satisfy

$$\lim_{t \to +\infty} \|y_i(t) - y_j(t)\| = 0$$

for all $i, j = 1, 2, ..., N$.

Remark 4.3 Recently, several authors have focused on the output synchronization problem of multi-agent systems, and several sufficient conditions have been presented to achieve output synchronization by designing appropriate control inputs [113–117]. In [113], a group of identical agents with nonlinear dynamics was considered. The authors investigated a system containing N nonidentical agents modeled by linear or nonlinear ordinary differential equations [114–117]. Nevertheless, only a few researchers have analyzed the output synchronization of CDNs with directed and undirected topologies.

4.3 Output Synchronization of CDNs with Directed and Undirected Topologies

In this section, we focus on the networks (4.1) with directed and undirected topologies and study their output synchronization. By applying the Lyapunov functional method and Barbalat's lemma, several output synchronization conditions are obtained.

4.3.1 Output Synchronization of a Complex Dynamical Network with a Directed Topology

Based on Lemma 1.10, a positive vector $\theta = (\theta_1, F, \theta_2, \cdots, \theta_N)^T \in \mathbb{R}^N$ satisfying $\theta^T G = 0$ can be easily found. Let $\xi_i = \frac{\theta_i}{\theta_1+\theta_2+\cdots+\theta_N}$ and $\bar{x}(t) = \sum_{i=1}^{N} \xi_i x_i(t)$. Then,

$$
\begin{aligned}
\dot{\bar{x}}(t) &= \sum_{i=1}^{N} \xi_i f(x_i(t)) + c \sum_{i=1}^{N} \xi_i \sum_{j=1}^{N} G_{ij} \Gamma x_j(t) \\
&= \sum_{i=1}^{N} \xi_i f(x_i(t)) + c \sum_{j=1}^{N} \left(\sum_{i=1}^{N} \xi_i G_{ij} \right) \Gamma x_j(t) \\
&= \sum_{i=1}^{N} \xi_i f(x_i(t)).
\end{aligned}
$$

By defining $e_i(t) = (e_{i1}(t), e_{i2}(t), \cdots, e_{in}(t))^T = x_i(t) - \bar{x}(t)(i = 1, 2, \cdots, N)$, one obtains

$$
\dot{e}_i(t) = f(x_i(t)) - \sum_{j=1}^{N} \xi_j f(x_j(t)) + c \sum_{j=1}^{N} G_{ij} \Gamma e_j(t), \tag{4.2}
$$

where $i = 1, 2, \cdots, N$.

Furthermore, nonsingular matrices $H \in \mathbb{R}^{n \times n}$ and $U \in \mathbb{R}^{q \times q}$ apparently exist such that

$$
C = U \begin{pmatrix} I_r & 0 \\ 0 & 0 \end{pmatrix} H,
$$

where $r = \text{Rank}(C)$. Therefore,

$$
\begin{cases}
\dot{e}_i(t) = f(x_i(t)) - \sum_{j=1}^{N} \xi_j f(x_j(t)) \\
\qquad\quad + c \sum_{j=1}^{N} G_{ij} \Gamma e_j(t), \\
y_i(t) = U \begin{pmatrix} I_r & 0 \\ 0 & 0 \end{pmatrix} H e_i(t) + C\bar{x}(t),
\end{cases}
$$

where $i = 1, 2, \cdots, N$. Quite evidently, the outputs of the network (4.1) achieve synchronization if and only if

$$
\lim_{t \to +\infty} \|z_i(t)\| = 0, \quad i = 1, 2, \cdots, N,
$$

where $z_i(t) = (z_{i1}(t), z_{i2}(t), \cdots, z_{ir}(t))^T = \hat{H} e_i(t) \in \mathbb{R}^r$, $\hat{H} = (I_r \;\; 0)_{r \times n} H$. That is,

$$
\begin{cases}
\dot{e}_i(t) = f(x_i(t)) - \sum_{j=1}^{N} \xi_j f(x_j(t)) \\
\qquad\quad + c \sum_{j=1}^{N} G_{ij} \Gamma e_j(t), \\
z_i(t) = \hat{H} e_i(t),
\end{cases} \tag{4.3}
$$

where $i = 1, 2, \cdots, N$.

Denote

$$\xi = \mathrm{diag}(\xi_1, \xi_2, \cdots, \xi_N),$$
$$e(t) = (e_1^T(t), e_2^T(t), \cdots, e_N^T(t))^T,$$
$$z(t) = (z_1^T(t), z_2^T(t), \cdots, z_N^T(t))^T.$$

Theorem 4.4 *If there exists a matrix* $0 < Q \in \mathbb{R}^{rN\times rN}$ *such that*

$$\begin{pmatrix} \mathbb{W} & I_N \otimes \hat{H}^T \\ I_N \otimes \hat{H} & -Q \end{pmatrix} \leqslant 0, \tag{4.4}$$

where $\mathbb{W} = (2\xi)\otimes(P\Delta - \eta I_n) + (c\xi G + cG^T\xi)\otimes(P\Gamma)$, *then the outputs of the network* (4.1) *achieve synchronization.*

Proof For system (4.3), the Lyapunov functional is given as follows:

$$V_1(e(t)) = \sum_{i=1}^{N} \xi_i e_i^T(t) P e_i(t). \tag{4.5}$$

By taking the derivative of $V_1(e(t))$, one obtains

$$\begin{aligned}\dot{V}_1(e(t)) &= 2\sum_{i=1}^{N} \xi_i e_i^T(t) P \dot{e}_i(t)\\ &= 2\sum_{i=1}^{N} \xi_i e_i^T(t) P[f(x_i(t)) - f(\bar{x}(t)) + f(\bar{x}(t))\\ &\quad -\sum_{j=1}^{N} \xi_j f(x_j(t)) + c\sum_{j=1}^{N} G_{ij}\Gamma e_j(t)].\end{aligned} \tag{4.6}$$

Obviously,

$$\begin{aligned}&\sum_{i=1}^{N} \xi_i e_i^T(t) P[f(x_i(t)) - f(\bar{x}(t))]\\ =&\sum_{i=1}^{N} \xi_i (x_i(t) - \bar{x}(t))^T P(f(x_i(t)) - f(\bar{x}(t)))\\ \leqslant&\sum_{i=1}^{N} \xi_i (x_i(t) - \bar{x}(t))^T (P\Delta - \eta I_n)(x_i(t) - \bar{x}(t))\end{aligned}$$

$$= \sum_{i=1}^{N} \xi_i e_i^T(t)(P\Delta - \eta I_n)e_i(t). \tag{4.7}$$

Furthermore,

$$\begin{aligned}
\sum_{i=1}^{N} \xi_i e_i(t) &= \sum_{i=1}^{N} \xi_i \left(x_i(t) - \sum_{j=1}^{N} \xi_j x_j(t) \right) \\
&= \sum_{i=1}^{N} \xi_i x_i(t) - \sum_{j=1}^{N} \left(\sum_{i=1}^{N} \xi_i \right) \xi_j x_j(t) \\
&= \sum_{i=1}^{N} \xi_i x_i(t) - \sum_{j=1}^{N} \xi_j x_j(t) \\
&= 0.
\end{aligned} \tag{4.8}$$

From (4.8), the following can be deduced:

$$\sum_{i=1}^{N} \xi_i e_i^T(t) P \left(f(\bar{x}(t)) - \sum_{j=1}^{N} \xi_j f(x_j(t)) \right) = 0. \tag{4.9}$$

Therefore, according to (4.4), (4.6), (4.7) and (4.9), one obtains

$$\begin{aligned}
\dot{V}_1(e(t)) &\leqslant 2 \sum_{i=1}^{N} \xi_i e_i^T(t)(P\Delta - \eta I_n)e_i(t) \\
&\quad + 2c \sum_{i=1}^{N} \sum_{j=1}^{N} \xi_i G_{ij} e_i^T(t) P \Gamma e_j(t) \\
&= e^T(t) \Big[(2\xi) \otimes (P\Delta - \eta I_n) \\
&\quad + (c\xi G + cG^T \xi) \otimes (P\Gamma) \Big] e(t) \\
&\leqslant -e^T(t)(I_N \otimes \hat{H}^T) Q^{-1} (I_N \otimes \hat{H}) e(t) \\
&= -e^T(t)(\hat{\xi} \otimes \hat{H}^T) \varXi (\hat{\xi} \otimes \hat{H}) e(t) \\
&\leqslant -\lambda_m(\varXi) e^T(t)(\hat{\xi} \otimes \hat{H}^T)(\hat{\xi} \otimes \hat{H}) e(t) \\
&= -\lambda_m(\varXi) e^T(t) \left[\xi \otimes (\hat{H}^T \hat{H}) \right] e(t)
\end{aligned}$$

$$= -\lambda_m(\Xi)\sum_{i=1}^{N}\xi_i e_i^T(t)\hat{H}^T\hat{H}e_i(t), \tag{4.10}$$

where $\Xi = (\hat{\xi}^{-1}\otimes I_r)Q^{-1}(\hat{\xi}^{-1}\otimes I_r)$, $\hat{\xi} = \text{diag}(\sqrt{\xi_1}, \sqrt{\xi_2}, \cdots, \sqrt{\xi_N})$.

Thus, it follows from (4.10) that $V_1(e(t))$ is nonincreasing, and $e_i(t), i = 1, 2, \cdots, N$ are bounded. Furthermore, because $V_1(e(t)) \geqslant 0$, it can be concluded that $\lim_{t\to+\infty} V_1(e(t))$ exists, and it can be inferred that $\lim_{t\to+\infty} V_1(e(t))$ is a nonnegative real number. That is,

$$\lim_{t\to+\infty} V_1(e(t)) = \sigma \in \mathbb{R} \geqslant 0. \tag{4.11}$$

Let $\alpha(t) = \sum_{i=1}^{N}\xi_i e_i^T(t)\hat{H}^T\hat{H}e_i(t)$. From (4.10) and (4.11), one obtains

$$\begin{aligned}\lim_{t\to+\infty}\int_0^t \alpha(s)ds &\leqslant -\frac{1}{\lambda_m(\Xi)}\lim_{t\to+\infty}\int_0^t \dot{V}_1(e(s))ds\\ &= \frac{1}{\lambda_m(\Xi)}\Big(V_1(e(0)) - \lim_{t\to+\infty} V_1(e(t))\Big)\\ &= \frac{1}{\lambda_m(\Xi)}\Big(V_1(e(0)) - \sigma\Big).\end{aligned} \tag{4.12}$$

Moreover,

$$\begin{aligned}\dot{\alpha}(t) &= 2\sum_{i=1}^{N}\xi_i e_i^T(t)\hat{H}^T\hat{H}\dot{e}_i(t)\\ &= 2\sum_{i=1}^{N}\xi_i e_i^T(t)\hat{H}^T\hat{H}\Big[f(x_i(t)) - f(\bar{x}(t)) + f(\bar{x}(t))\\ &\quad -\sum_{j=1}^{N}\xi_j f(x_j(t)) + c\sum_{j=1}^{N}G_{ij}\Gamma e_j(t)\Big]\\ &= 2\sum_{i=1}^{N}\xi_i e_i^T(t)\hat{H}^T\hat{H}\Big[f(x_i(t)) - f(\bar{x}(t)) + c\sum_{j=1}^{N}G_{ij}\Gamma e_j(t)\Big].\end{aligned} \tag{4.13}$$

Because $e_i(t), i = 1, 2, \cdots, N$ are bounded,

$$\|e_i(t)\| \leqslant \kappa_i \in \mathbb{R} > 0 \tag{4.14}$$

for any $t \in [0, +\infty)$.

Because $f(\cdot)$ satisfies the Lipschitz condition (see Remark 4.1), then there exists $0 < \gamma \in \mathbb{R}$ such that

$$\|f(x_i(t)) - f(\bar{x}(t))\| \leqslant \gamma \|x_i(t) - \bar{x}(t)\| = \gamma \|e_i(t)\|, \tag{4.15}$$

where $i = 1, 2, \cdots, N$.

According to (4.13), (4.14), (4.15), one obtains

$$\begin{aligned}
|\dot{\alpha}(t)| &= \left|2\sum_{i=1}^{N} \xi_i e_i^T(t)\hat{H}^T\hat{H}\Big[f(x_i(t)) - f(\bar{x}(t)) + c\sum_{j=1}^{N} G_{ij}\Gamma e_j(t)\Big]\right| \\
&\leqslant 2\sum_{i=1}^{N} \xi_i \left|e_i^T(t)\hat{H}^T\hat{H}\left[f(x_i(t)) - f(\bar{x}(t))\right]\right| \\
&\quad +2c\sum_{i=1}^{N}\sum_{j=1}^{N} \xi_i |G_{ij}| \left|e_i^T(t)\hat{H}^T\hat{H}\Gamma e_j(t)\right| \\
&\leqslant \sum_{i=1}^{N} \xi_i e_i^T(t)\left(\hat{H}^T\hat{H}\right)^2 e_i(t) \\
&\quad +\sum_{i=1}^{N} \xi_i \left[f(x_i(t)) - f(\bar{x}(t))\right]^T \left[f(x_i(t)) - f(\bar{x}(t))\right] \\
&\quad +c\sum_{i=1}^{N}\sum_{j=1}^{N} \xi_i |G_{ij}| e_i^T(t)\left(\hat{H}^T\hat{H}\right)^2 e_i(t) \\
&\quad +c\sum_{i=1}^{N}\sum_{j=1}^{N} \xi_i |G_{ij}| e_j^T(t)\Gamma^2 e_j(t) \\
&\leqslant \sum_{i=1}^{N} \xi_i \left(\lambda_M^2(\hat{H}^T\hat{H}) + \gamma^2\right)\|e_i(t)\|^2 \\
&\quad +c\sum_{i=1}^{N}\sum_{j=1}^{N} \xi_i |G_{ij}| \left(\lambda_M^2(\hat{H}^T\hat{H})\|e_i(t)\|^2 + \lambda_M^2(\Gamma)\|e_j(t)\|^2\right) \\
&\leqslant c\sum_{i=1}^{N}\sum_{j=1}^{N} \xi_i |G_{ij}| \left(\lambda_M^2(\hat{H}^T\hat{H})\kappa_i^2 + \lambda_M^2(\Gamma)\kappa_j^2\right) \\
&\quad +\sum_{i=1}^{N} \xi_i \left(\lambda_M^2(\hat{H}^T\hat{H}) + \gamma^2\right)\kappa_i^2.
\end{aligned} \tag{4.16}$$

Obviously, $\alpha(t)$ is uniformly continuous. From Lemma 1.8, one obtains

$$\lim_{t\to+\infty} \alpha(t) = 0. \tag{4.17}$$

By applying (4.17), one obtains

$$\lim_{t\to+\infty} \|z_i(t)\| = \lim_{t\to+\infty} \|\hat{H}e_i(t)\| = 0,$$

where $i = 1, 2, \cdots, N$. Therefore, the outputs of the network model (4.1) achieve synchronization. □

Remark 4.5 In the last few years, studies have been carried out on the output synchronization of multi-agent systems [33, 35, 98, 105, 106]. In [16–19], the authors focused on multi-agent systems, and investigated passivity-based control for the output synchronization problem. By using the passivity property of each agent and the designed control inputs, some conditions were proposed to attain output synchronization. Based on assumptions about the agents and the topology of the network, Grip et al. [106] designed decentralized controllers to achieve output synchronization. In this section, the synchronization of outputs is analyzed for a complex dynamical network (4.1) with a directed topology by using Barbalat's lemma. Obviously, this work is totally different from the above-mentioned studies [33, 35, 98, 105, 106].

4.3.2 Output Synchronization of a Complex Dynamical Network with an Undirected Topology

This subsection discusses a case in which G is symmetric. Under this circumstance, the following can be easily derived:

$$\mathbf{1}_N^T G = 0,$$

where $\mathbf{1}_N = (1, 1, \cdots, 1)^T \in \mathbb{R}^N$.

Let $\bar{x}(t) = \frac{1}{N}\sum_{i=1}^N x_i(t)$. Then,

$$\dot{\bar{x}}(t) = \frac{1}{N}\sum_{i=1}^N f(x_i(t)).$$

Define $e_i(t) = (e_{i1}(t), e_{i2}(t), \cdots, e_{in}(t))^T = x_i(t) - \bar{x}(t) (i = 1, 2, \cdots, N)$. Then,

$$\dot{e}_i(t) = f(x_i(t)) - \frac{1}{N}\sum_{j=1}^{N} f(x_j(t)) + c\sum_{j=1}^{N} G_{ij}\Gamma e_j(t), \tag{4.18}$$

where $i = 1, 2, \cdots, N$.

Similarly, it is clear that a complex network (4.1) achieves output synchronization if and only if

$$\lim_{t\to+\infty} \|z_i(t)\| = 0, \quad i = 1, 2, \cdots, N,$$

where $z_i(t) = (z_{i1}(t), z_{i2}(t), \cdots, z_{ir}(t))^T = \hat{H}e_i(t) \in \mathbb{R}^r$, $\hat{H} = (I_r \;\; 0)_{r\times n}H$, H has the same meaning as in (4.3), that is,

$$\begin{cases} \dot{e}_i(t) = f(x_i(t)) - \frac{1}{N}\sum_{j=1}^{N} f(x_j(t)) \\ \qquad\quad +c\sum_{j=1}^{N} G_{ij}\Gamma e_j(t), \\ z_i(t) = \hat{H}e_i(t), \quad i = 1, 2, \cdots, N. \end{cases} \tag{4.19}$$

Denote

$$e(t) = (e_1^T(t), e_2^T(t), \cdots, e_N^T(t))^T,$$
$$z(t) = (z_1^T(t), z_2^T(t), \cdots, z_N^T(t))^T.$$

Theorem 4.6 *If there exists a matrix* $0 < Q \in \mathbb{R}^{rN\times rN}$ *such that*

$$\begin{pmatrix} \hat{\mathbb{W}} & I_N \otimes \hat{H}^T \\ I_N \otimes \hat{H} & -Q \end{pmatrix} \leqslant 0, \tag{4.20}$$

where $\hat{\mathbb{W}} = I_N \otimes (P\Delta - \eta I_n) + (cG) \otimes (P\Gamma)$, *then the outputs of the network* (4.1) *achieve synchronization.*

Proof The following Lyapunov functional is defined for the system (4.19):

$$V_2(e(t)) = \frac{1}{2}\sum_{i=1}^{N} e_i^T(t)Pe_i(t). \tag{4.21}$$

Then, the following is obtained:

$$\begin{aligned} \dot{V}_2(e(t)) &= \sum_{i=1}^{N} e_i^T(t)P\dot{e}_i(t) \\ &= \sum_{i=1}^{N} e_i^T(t)P\Big[f(x_i(t)) - f(\bar{x}(t)) + f(\bar{x}(t)) \end{aligned}$$

$$
\begin{aligned}
&-\frac{1}{N}\sum_{j=1}^{N} f(x_j(t)) + c\sum_{j=1}^{N} G_{ij}\Gamma e_j(t)\Big] \\
&\leqslant \sum_{i=1}^{N} e_i^T(t)(P\Delta - \eta I_n)e_i(t) + c\sum_{i=1}^{N}\sum_{j=1}^{N} G_{ij}e_i^T(t)P\Gamma e_j(t) \\
&= e^T(t)\left[I_N \otimes (P\Delta - \eta I_n) + (cG)\otimes(P\Gamma)\right]e(t) \\
&\leqslant -e^T(t)(I_N \otimes \hat{H}^T)Q^{-1}(I_N \otimes \hat{H})e(t).
\end{aligned}
$$

By applying the same arguments as Theorem 4.4, one obtains $\lim_{t\to+\infty}\|z_i(t)\| = 0, i = 1, 2, \cdots, N$. Hence, the outputs of the network model (4.1) achieve synchronization. □

Remark 4.7 In [24], the authors focused on an output coupled complex network model and studied its output synchronization. Furthermore, some sufficient conditions to ensure local and global exponential output synchronization were established. However, these results were derived based on a particular solution of an isolate node of the network, which is probably hard to obtain in some engineering applications. This restriction is removed in the present work.

4.4 Adaptive Output Synchronization of CDNs with Directed and Undirected Topologies

In the past 10 years, adaptive control has been developed as an effective method to design control inputs for agents [118–120]. Practically, the adaptive control strategy can also be applied to adjust the coupling weights of network models. This section focuses on the adaptive output synchronization of a complex dynamical network (4.1) with, respectively, directed and undirected topologies.

4.4.1 Adaptive Output Synchronization of a Complex Dynamical Network with a Directed Topology

The main goal of this part of the research is to design an adaptive strategy to adjust the coupling weights of a complex dynamical network (4.1) to achieve output synchronization. Toward this end, the network model (4.1) can be described as follows:

$$
\begin{cases}
\dot{x}_i(t) = f(x_i(t)) + c\sum_{j=1}^{N} G_{ij}(t)\Gamma x_j(t), \\
y_i(t) = \quad Cx_i(t), \quad t \geqslant 0,
\end{cases}
\tag{4.22}
$$

where $i = 1, 2, \cdots, N$. $G(0) = (G_{ij}(0))_{N\times N}$ is a constant matrix. If the connection from nodes i to j at time 0 is existent, then $G_{ij}(0) > 0$; if the connection from nodes i to j at time 0 is non-existent, then $G_{ij}(0) = 0 (i \neq j)$. Moreover, $G_{ii}(0)$ satisfies

$$G_{ii}(0) = -\sum_{\substack{j=1 \\ j\neq i}}^{N} G_{ij}(0), \quad i = 1, 2, \cdots, N.$$

Remark 4.8 In this subsection, the network (4.22) is considered to be connected and has a fixed topology. That is, if the connection from nodes i to $j (i \neq j)$ at time 0 is non-existent, then $G_{ij}(t) = 0$ for all $t > 0$; if the connection from nodes i to $j (i \neq j)$ at time 0 is existent, then $G_{ij}(t) > 0$ for all $t > 0$.

Based on Lemma 1.10, a positive vector $\theta = (\theta_1, \theta_2, \cdots, \theta_N)^T \in \mathbb{R}^N$ can be found, which satisfies $\theta^T G(0) = 0$ and

$$\sum_{j=1}^{N} \hat{G}_{ij} = \sum_{j=1}^{N} \hat{G}_{ji} = 0 \text{ for all } i = 1, 2, ..., N,$$

where $\Xi = \text{diag}(\theta_1, \theta_2, \cdots, \theta_N)$, $\hat{G} = \Xi G(0) + G^T(0)\Xi$.

Let $\xi_i = \frac{\theta_i}{\theta_1+\theta_2+\cdots+\theta_N}$, $i = 1, 2, ..., N$. Then, $(\xi_1, \xi_2, \cdots, \xi_N)G(0) = 0$ and

$$G_{ii}(0) = -\frac{\sum_{\substack{j=1 \\ j\neq i}}^{N} \left(\xi_i G_{ij}(0) + \xi_j G_{ji}(0)\right)}{2\xi_i} \tag{4.23}$$

for all $i = 1, 2, ..., N$.

Use the following adaptive law:

$$\dot{G}_{ij}(t) = \begin{cases} \beta_{ij}\xi_j (x_i(t) - x_j(t))^T P\Gamma (x_i(t) - x_j(t)), \\ \qquad\qquad \text{if } (i, j) \in \mathcal{E} \text{ and } (j, i) \in \mathcal{E}, \\ -\sum_{\substack{l=1 \\ l\neq i}}^{N} \dot{G}_{il}(t), \qquad\qquad \text{if } i = j, \\ 0, \qquad\qquad\qquad\qquad \text{otherwise}, \end{cases} \tag{4.24}$$

where $\beta_{ij} = \beta_{ji}$ are positive constants.

By applying (4.24), one obtains

$$\xi_j \dot{G}_{ji}(t) = \xi_i \dot{G}_{ij}(t), \tag{4.25}$$

$$\sum_{j=1}^{N} \dot{G}_{ij}(t) = 0,$$

where $i, j = 1, 2, ..., N$.

Thus, one obtains

$$\sum_{j=1}^{N} \xi_j \dot{G}_{ji}(t) = \sum_{j=1}^{N} \xi_i \dot{G}_{ij}(t)$$
$$= \xi_i \left(\sum_{j=1}^{N} \dot{G}_{ij}(t) \right)$$
$$= 0,$$

$$\dot{G}_{ii}(t) = -\frac{\sum_{\substack{j=1 \\ j \neq i}}^{N} \left(\xi_i \dot{G}_{ij}(t) + \xi_j \dot{G}_{ji}(t) \right)}{2\xi_i}. \tag{4.26}$$

Because $(\xi_1, \xi_2, \cdots, \xi_N)G(0) = 0$,

$$\sum_{j=1}^{N} \xi_j G_{ji}(t) = 0$$

for all t and $i \in \{1, 2, ..., N\}$.

According to (4.23), (4.24), (4.26), one obtains

$$G_{ii}(t) = -\frac{\sum_{\substack{j=1 \\ j \neq i}}^{N} \left(\xi_i G_{ij}(t) + \xi_j G_{ji}(t) \right)}{2\xi_i} \tag{4.27}$$

for all t and $i \in \{1, 2, ..., N\}$.

Let $\bar{x}(t) = \sum_{i=1}^{N} \xi_i x_i(t)$. Then,

$$\dot{\bar{x}}(t) = \sum_{i=1}^{N} \xi_i f(x_i(t)) + c \sum_{j=1}^{N} \left(\sum_{i=1}^{N} \xi_i G_{ij}(t) \right) \Gamma x_j(t)$$
$$= \sum_{i=1}^{N} \xi_i f(x_i(t)).$$

Define $e_i(t) = (e_{i1}(t), e_{i2}(t), \cdots, e_{in}(t))^T = x_i(t) - \bar{x}(t), i = 1, 2, \cdots, N$. Then, the following equation holds:

$$\dot{e}_i(t) = f(x_i(t)) - \sum_{j=1}^{N} \xi_j f(x_j(t)) + c \sum_{j=1}^{N} G_{ij}(t) \Gamma e_j(t), \tag{4.28}$$

where $i = 1, 2, \cdots, N$.

Similarly, it is clear that the complex network (4.22) achieves output synchronization if and only if

$$\lim_{t\to+\infty} \|z_i(t)\| = 0, \quad i = 1, 2, \cdots, N,$$

where $z_i(t) = (z_{i1}(t), z_{i2}(t), \cdots, z_{ir}(t))^T = \hat{H}e_i(t) \in \mathbb{R}^r$, $\hat{H} = (I_r \;\; 0)_{r\times n}H$, H has the same meaning as in (4.3). That is,

$$\begin{cases} \dot{e}_i(t) &= f(x_i(t)) - \sum_{j=1}^N \xi_j f(x_j(t)) \\ &\quad +c\sum_{j=1}^N G_{ij}(t)\Gamma e_j(t), \\ \dot{G}_{ij}(t) &= \beta_{ij}\xi_j(e_i(t)-e_j(t))^T P\Gamma(e_i(t) \\ &\quad - e_j(t)),\ (i,j)\in\mathcal{E} \text{ and } (j,i)\in\mathcal{E}, \\ \dot{G}_{ii}(t) &= -\sum_{\substack{j=1\\ j\neq i}}^N \dot{G}_{ij}(t), \\ z_i(t) &= \hat{H}e_i(t),\ i=1,2,\cdots,N. \end{cases} \tag{4.29}$$

Denote

$$A_{ij}(t) = \begin{cases} \xi_i G_{ij}(t) + \xi_j G_{ji}(t), \text{ if } (i,j) \text{ and } (j,i)\in\mathcal{E}, \\ -\sum_{\substack{l=1\\ l\neq i}}^N A_{il}(t), \qquad \text{if } i=j, \\ 0, \qquad \text{otherwise}, \end{cases}$$

$$M_{ij} = \begin{cases} \xi_i G_{ij}(0), \text{ if } (i,j)\in\mathcal{E} \text{ and } (j,i)\notin\mathcal{E}, \\ \xi_j G_{ji}(0), \text{ if } (j,i)\in\mathcal{E} \text{ and } (i,j)\notin\mathcal{E}, \\ -\sum_{\substack{l=1\\ l\neq i}}^N M_{il}, \qquad \text{if } i=j, \\ 0, \qquad \text{otherwise}. \end{cases}$$

Based on the definitions of $A_{ij}(t)$ and M_{ij},

$$\xi_i G_{ij}(t) + \xi_j G_{ji}(t) = A_{ij}(t) + M_{ij}, \quad i,j\in\{1,2,\cdots,N\}.$$

Theorem 4.9 *The outputs of the network model* (4.22) *achieve synchronization based on the adaptive law* (4.24) *if there exist matrices* $0 < Q \in \mathbb{R}^{rN\times rN}$ *and* $\kappa = (\kappa_{ij})_{N\times N}$ *such that*

$$\begin{pmatrix} \mathbb{W}_1 & I_N\otimes\hat{H}^T \\ I_N\otimes\hat{H} & -Q \end{pmatrix} \leqslant 0, \tag{4.30}$$

where $\xi = \text{diag}(\xi_1, \xi_2, \cdots, \xi_N)$, $M = (M_{ij})_{N\times N}$, $\mathbb{W}_1 = (2\xi)\otimes(P\Delta - \eta I_n) + [c(M+\kappa)]\otimes(P\Gamma)$, $\kappa_{ii} = -\sum_{\substack{j=1\\ j\neq i}}^N \kappa_{ij}$, *and* $\kappa_{ij} = \kappa_{ji} (i\neq j)$ *are nonnegative constants, and* $\kappa_{ij} = 0 (i\neq j)$ *if and only if* $A_{ij}(t) = 0$.

Proof The Lyapunov functional for the system (4.29) is given as follows:

$$V_3(\hat{e}(t)) = \sum_{i=1}^{N} \xi_i e_i^T(t) P e_i(t) + c\sum_{i=1}^{N} \sum_{\substack{(i,j)\in\mathcal{E},\\(j,i)\in\mathcal{E}}} \frac{(\xi_i G_{ij}(t) + \xi_j G_{ji}(t) - \kappa_{ij})^2}{8\beta_{ij}\xi_i\xi_j}, \tag{4.31}$$

where $e(t) = (e_1^T(t), e_2^T(t), \cdots, e_N^T(t))^T$, $\hat{e}(t) = (e^T(t), \underbrace{\cdots, G_{ij}(t), \cdots}_{(i,j) \text{ and } (j,i)\in\mathcal{E}})^T$.

By calculating the time derivation of $V_3(\hat{e}(t))$ along the trajectory of the system (4.29), one obtains

$$\begin{aligned}
\dot{V}_3(\hat{e}(t)) &= 2\sum_{i=1}^{N} \xi_i e_i^T(t) P \dot{e}_i(t) \\
&\quad + \sum_{i=1}^{N} \sum_{\substack{(i,j)\in\mathcal{E},\\(j,i)\in\mathcal{E}}} \frac{c}{4\beta_{ij}\xi_i\xi_j}(\xi_i G_{ij}(t) + \xi_j G_{ji}(t) \\
&\quad - \kappa_{ij})(\xi_i \dot{G}_{ij}(t) + \xi_j \dot{G}_{ji}(t)) \\
&= 2\sum_{i=1}^{N} \xi_i e_i^T(t) P[f(x_i(t)) - f(\bar{x}(t)) + f(\bar{x}(t)) \\
&\quad - \sum_{j=1}^{N} \xi_j f(x_j(t)) + c\sum_{j=1}^{N} G_{ij}(t)\Gamma e_j(t)] \\
&\quad + \frac{c}{2}\sum_{i=1}^{N} \sum_{\substack{(i,j)\in\mathcal{E},\\(j,i)\in\mathcal{E}}} (\xi_i G_{ij}(t) + \xi_j G_{ji}(t) - \kappa_{ij})(e_i(t) \\
&\quad - e_j(t))^T P\Gamma(e_i(t) - e_j(t)) \\
&\leqslant \sum_{i=1}^{N} 2\xi_i e_i^T(t)(P\Delta - \eta I_n)e_i(t) \\
&\quad + 2c\sum_{i=1}^{N}\sum_{j=1}^{N} \xi_i G_{ij}(t) e_i^T(t) P\Gamma e_j(t)
\end{aligned}$$

$$+\frac{c}{2}\sum_{i=1}^{N}\sum_{\substack{(i,j)\in\mathcal{E},\\(j,i)\in\mathcal{E}}}(\xi_i G_{ij}(t)+\xi_j G_{ji}(t)-\kappa_{ij})(e_i(t)$$
$$-e_j(t))^T P\Gamma(e_i(t)-e_j(t)). \tag{4.32}$$

The following can thus be easily derived:

$$\sum_{i=1}^{N}\sum_{\substack{(i,j)\in\mathcal{E},\\(j,i)\in\mathcal{E}}}(\xi_i G_{ij}(t)+\xi_j G_{ji}(t)-\kappa_{ij})(e_i(t)$$
$$-e_j(t))^T P\Gamma(e_i(t)-e_j(t))$$
$$=-2\sum_{i=1}^{N}\sum_{j=1}^{N}(A_{ij}(t)-\kappa_{ij})e_i^T(t)P\Gamma e_j(t). \tag{4.33}$$

Based on (4.30), (4.32), (4.33), one obtains

$$\begin{aligned}
\dot{V}_3(\hat{e}(t)) \leqslant &\sum_{i=1}^{N}2\xi_i e_i^T(t)(P\Delta-\eta I_n)e_i(t)\\
&+2c\sum_{i=1}^{N}\sum_{j=1}^{N}\xi_i G_{ij}(t)e_i^T(t)P\Gamma e_j(t)\\
&-c\sum_{i=1}^{N}\sum_{j=1}^{N}(A_{ij}(t)-\kappa_{ij})e_i^T(t)P\Gamma e_j(t)\\
=&\sum_{i=1}^{N}2\xi_i e_i^T(t)(P\Delta-\eta I_n)e_i(t)\\
&+c\sum_{i=1}^{N}\sum_{j=1}^{N}(\xi_i G_{ij}(t)+\xi_j G_{ji}(t))e_i^T(t)P\Gamma e_j(t)\\
&-c\sum_{i=1}^{N}\sum_{j=1}^{N}(A_{ij}(t)-\kappa_{ij})e_i^T(t)P\Gamma e_j(t)\\
=&\sum_{i=1}^{N}2\xi_i e_i^T(t)(P\Delta-\eta I_n)e_i(t)\\
&+c\sum_{i=1}^{N}\sum_{j=1}^{N}(M_{ij}+\kappa_{ij})e_i^T(t)P\Gamma e_j(t)
\end{aligned}$$

$$
\begin{aligned}
&= e^T(t)\Big\{(2\xi)\otimes(P\Delta-\eta I_n)+[c(M+\kappa)]\otimes(P\Gamma)\Big\}e(t)\\
&\leqslant -e^T(t)(I_N\otimes\hat{H}^T)Q^{-1}(I_N\otimes\hat{H})e(t)\\
&\leqslant -\lambda_m(\Xi)\sum_{i=1}^{N}\xi_i e_i^T(t)\hat{H}^T\hat{H}e_i(t),
\end{aligned}
\tag{4.34}
$$

where $\Xi = (\hat{\xi}^{-1}\otimes I_r)Q^{-1}(\hat{\xi}^{-1}\otimes I_r)$, and $\hat{\xi} = \mathrm{diag}(\sqrt{\xi_1}, \sqrt{\xi_2}, \cdots, \sqrt{\xi_N})$. Obviously, $V_3(\hat{e}(t))$ is nonincreasing. Therefore, $e_i(t)(i = 1, 2, \cdots, N)$ and the coupling weights $G_{ij}(t)((i,j)$ and $(j,i) \in \mathcal{E})$ are bounded. Because $G_{ij}(t)((i,j)$ and $(j,i) \in \mathcal{E})$ are monotonically increasing (see (4.24)), one obtains

$$
\lim_{t\to+\infty} G_{ij}(t) = G_{ij}^0 \in \mathbb{R} > 0,\ (i,j) \text{ and } (j,i)\in\mathcal{E}.
$$

Let $\alpha(t) = \sum_{i=1}^{N}\xi_i e_i^T(t)\hat{H}^T\hat{H}e_i(t)$. Then, by applying the same arguments as in Theorem 4.4, one obtains

$$
\lim_{t\to+\infty}\alpha(t) = 0. \tag{4.35}
$$

From (4.35), we have

$$
\lim_{t\to+\infty}\|z_i(t)\| = 0,\ i = 1, 2, \cdots, N.
$$

Thus, the outputs of the network model (4.22) achieve synchronization based on the adaptive law (4.24).□

A design procedure of adaptive law is introduced with the help of the above analysis.

Procedure 1:

Step 1: Choose the appropriate η, P, and Δ such that $f(\cdot)$ satisfies the assumption (A1).

Step 2: Find the positive vector $\theta = (\theta_1, \theta_2, \cdots, \theta_N)^T \in \mathbb{R}^N$ satisfying $\theta^T G(0) = 0$.

Step 3: Calculate $\hat{H}$, ξ_i, and $M_{ij}(i, j = 1, 2, \cdots, N)$.

Step 4: Solve the linear matrix inequality (LMI) (4.30). If a feasible solution exists, go to Step 5.

Step 5: Choose the appropriate positive constants β_{ij} in (4.24).

Remark 4.10 For many network models, such as biological systems and neural networks, the use of adaptively adjusted coupling weights is more reasonable [85, 108]. In addition, the appropriate coupling weights can evidently improve the synchronizability of the networks. That the outputs of CDNs with a directed topology and adaptive coupling weights can reach synchronization is known to us; however, this subject has not been studied by researchers. In Theorem 4.9, a sufficient condition is constructed to enable the outputs of the network model (4.22) to achieve synchronization by applying the designed adaptive law. With the help of the YALMIP toolbox, Q and κ can be found to satisfy (4.30).

4.4.2 Adaptive Output Synchronization of a Complex Dynamical Network with an Undirected Topology

In this subsection, $G(t)$ is considered to be symmetric in the network (4.22). Obviously, the following can be derived:

$$\begin{cases} \dot{x}_i(t) = f(x_i(t)) + c\sum_{j=1}^{N} G_{ij}(t)\Gamma x_j(t), \\ y_i(t) = \quad Cx_i(t), \quad t \geqslant 0, \end{cases} \tag{4.36}$$

where $i = 1, 2, \cdots, N$, $G(t) = (G_{ij}(t))_{N\times N}$ is a time-varying matrix. If nodes i and j are connected at time t, then $G_{ij}(t) = G_{ji}(t) > 0$; otherwise, $G_{ij}(t) = G_{ji}(t) = 0 (i \neq j)$, and $G_{ii}(t)$ is defined by

$$G_{ii}(t) = -\sum_{\substack{j=1\\ j\neq i}}^{N} G_{ij}(t).$$

In this subsection, the network (4.36) is considered to be connected and has a fixed topology. That is, if nodes i and $j (i \neq j)$ are not connected, $G_{ij}(t) = 0$ for all t.

Use the following adaptive law:

$$\dot{G}_{ij}(t) = \begin{cases} \beta_{ij}(x_i(t) - x_j(t))^T P\Gamma(x_i(t) - x_j(t)), \\ \qquad\qquad\qquad\qquad\qquad \text{if } (i, j) \in \mathcal{E}, \\ -\sum_{\substack{l=1\\ l\neq i}}^{N} \dot{G}_{il}(t), \qquad\qquad \text{if } i = j, \\ 0, \qquad\qquad\qquad\qquad\quad \text{otherwise}, \end{cases} \tag{4.37}$$

where $\beta_{ij} = \beta_{ji}$ are positive constants.

Let $\bar{x}(t) = \frac{1}{N}\sum_{i=1}^{N} x_i(t)$. One obtains

$$\dot{\bar{x}}(t) = \frac{1}{N}\sum_{i=1}^{N} f(x_i(t)).$$

Take $e_i(t) = (e_{i1}(t), e_{i2}(t), \cdots, e_{in}(t))^T = x_i(t) - \bar{x}(t)$. Then, the dynamics of $e_i(t)$ is defined as:

$$\dot{e}_i(t) = f(x_i(t)) - \frac{1}{N}\sum_{j=1}^{N} f(x_j(t)) + c\sum_{j=1}^{N} G_{ij}(t)\Gamma e_j(t),$$

where $i = 1, 2, \cdots, N$.

Similarly, it is clear that the complex network (4.36) achieves output synchronization if and only if

$$\lim_{t\to+\infty} \|z_i(t)\| = 0, \quad i = 1, 2, \cdots, N,$$

where $z_i(t) = (z_{i1}(t), z_{i2}(t), \cdots, z_{ir}(t))^T = \hat{H}e_i(t) \in \mathbb{R}^r$, $\hat{H} = (I_r \;\; 0)_{r\times n}H$, H has the same meaning as in (4.3). That is,

$$\begin{cases} \dot{e}_i(t) \;=\; f(x_i(t)) + c\sum_{j=1}^{N} G_{ij}(t)\Gamma e_j(t) \\ \qquad\qquad -\frac{1}{N}\sum_{j=1}^{N} f(x_j(t)), \\ \dot{G}_{ij}(t) = \beta_{ij}(e_i(t) - e_j(t))^T P\Gamma(e_i(t) - e_j(t)), \\ \qquad\qquad\qquad\qquad (i, j) \in \mathcal{E}, \\ z_i(t) \;=\; \hat{H}e_i(t), \quad i = 1, 2, \cdots, N. \end{cases} \tag{4.38}$$

Theorem 4.11 *The outputs of the network model* (4.36) *achieve synchronization based on the following adaptive law:*

$$\dot{G}_{ij}(t) = \beta_{ij}(x_i(t) - x_j(t))^T P\Gamma(x_i(t) - x_j(t)),$$

$(i, j) \in \mathcal{E}$, where $\beta_{ij} = \beta_{ji} > 0$.

Proof The Lyapunov functional for the system (4.38) is given as follows:

$$V_4(\hat{e}(t)) = \sum_{i=1}^{N} e_i^T(t)Pe_i(t) + c\sum_{i=1}^{N}\sum_{(i,j)\in\mathcal{E}} \frac{(G_{ij}(t) - \hat{\kappa}_{ij})^2}{2\beta_{ij}}, \tag{4.39}$$

where $e(t) = (e_1^T(t), e_2^T(t), \cdots, e_N^T(t))^T$, $\hat{e}(t) = (e^T(t), \underbrace{\cdots, G_{ij}(t), \cdots}_{(i,j)\in\mathcal{E}})^T$, $\mathbb{R} \ni \hat{\kappa}_{ij} = \hat{\kappa}_{ji}(i \neq j) \geqslant 0$, and $\hat{\kappa}_{ij} = 0(i \neq j)$ is equivalent to $G_{ij}(t) = 0$.

Then,

$$\dot{V}_4(\hat{e}(t)) = 2\sum_{i=1}^{N} e_i^T(t)P\dot{e}_i(t) + c\sum_{i=1}^{N}\sum_{(i,j)\in\mathcal{E}} (G_{ij}(t) \\ -\hat{\kappa}_{ij})(e_i(t) - e_j(t))^T P\Gamma(e_i(t) - e_j(t))$$

$$
\begin{aligned}
&= 2\sum_{i=1}^{N} e_i^T(t)P[f(x_i(t)) - f(\bar{x}(t)) + f(\bar{x}(t)) \\
&\quad -\frac{1}{N}\sum_{j=1}^{N} f(x_j(t)) + c\sum_{j=1}^{N} G_{ij}(t)\Gamma e_j(t)] \\
&\quad -2c\sum_{i=1}^{N}\sum_{j=1}^{N}(G_{ij}(t) - \hat{\kappa}_{ij})e_i^T(t)P\Gamma e_j(t) \\
&\leqslant \sum_{i=1}^{N} 2e_i^T(t)(P\Delta - \eta I_n)e_i(t) \\
&\quad +2c\sum_{i=1}^{N}\sum_{j=1}^{N}\hat{\kappa}_{ij}e_i^T(t)P\Gamma e_j(t) \\
&= 2e^T(t)[I_N \otimes (P\Delta - \eta I_n) + (c\hat{\kappa}) \otimes (P\Gamma)]e(t),
\end{aligned}
$$

where $\hat{\kappa}_{ii} = -\sum_{\substack{j=1\\ j\neq i}}^{N} \hat{\kappa}_{ij}$, $\hat{\kappa} = (\hat{\kappa}_{ij})_{N\times N}$.

Obviously, there exists a unitary matrix $\phi = (\phi_1, \phi_2, \cdots, \phi_N) \in \mathbb{R}^{N\times N}$ such that $\phi^T\hat{\kappa}\phi = \Lambda$ with $\Lambda = \text{diag}(\lambda_1, \lambda_2, \cdots, \lambda_N)$. $\lambda_i, i = 1, 2, \cdots, N$, are the eigenvalues of $\hat{\kappa}$ and

$$
0 = \lambda_1 > \lambda_2 \geqslant \lambda_3 \geqslant \cdots \geqslant \lambda_N.
$$

Let $\eta(t) = (\eta_1^T(t), \eta_2^T(t), \cdots, \eta_N^T(t))^T = (\phi^T \otimes I_n)e(t)$. Because $\phi_1 = \frac{1}{\sqrt{N}}(1, 1, \cdots, 1)^T$, one obtains $\eta_1(t) = (\phi_1^T \otimes I_n)e(t) = 0$. Thus,

$$
\begin{aligned}
\dot{V}_4(\hat{e}(t)) &\leqslant 2e^T(t)\{I_N \otimes (P\Delta - \eta I_n) \\
&\quad +c(\phi \otimes I_n)[\Lambda \otimes (P\Gamma)](\phi^T \otimes I_n)\}e(t) \\
&= 2e^T(t)[I_N \otimes (P\Delta - \eta I_n)]e(t) \\
&\quad +2c\eta^T(t)[\Lambda \otimes (P\Gamma)]\eta(t) \\
&\leqslant 2e^T(t)[I_N \otimes (P\Delta - \eta I_n)]e(t) \\
&\quad +2c\lambda_2\eta^T(t)[I_N \otimes (P\Gamma)]\eta(t) \\
&= 2e^T(t)[I_N \otimes (P\Delta - \eta I_n)]e(t) \\
&\quad +2c\lambda_2 e^T(t)(\phi \otimes I_n)[I_N \otimes (P\Gamma)](\phi^T \otimes I_n)e(t) \\
&= 2e^T(t)\{I_N \otimes (P\Delta - \eta I_n) + c\lambda_2[I_N \otimes (P\Gamma)]\}e(t).
\end{aligned}
$$

By choosing a large enough $\hat{\kappa}_{ij}$ such that

$$\lambda_M(P\Delta - \eta I_n) + c\lambda_2\lambda_m(P\Gamma) + \frac{\lambda_M(\hat{H}^T\hat{H})}{2} \leqslant 0,$$

one obtains

$$\begin{aligned} &I_N \otimes (P\Delta - \eta I_n) + c\lambda_2[I_N \otimes (P\Gamma)] \\ &\leqslant -\frac{(I_N \otimes \hat{H}^T)(I_N \otimes \hat{H})}{2}. \end{aligned}$$

Therefore,

$$\dot{V}_4(\hat{e}(t)) \leqslant -e^T(t)(I_N \otimes \hat{H}^T)(I_N \otimes \hat{H})e(t). \tag{4.40}$$

By applying the same arguments as in Theorem 4.9, one obtains

$$\lim_{t\to+\infty} \|z_i(t)\| = 0,$$

where $i = 1, 2, \cdots, N$.

Thus, the outputs of the network model (4.36) achieve synchronization based on the adaptive law (4.37). □

Remark 4.12 In the past few decades, complex network models with known coupling weights have been widely applied in modeling complex systems in science, biology, and engineering, among other fields. Examples include networks of Lorenz oscillators, coupled neural networks, coupled Chua's circuits, linearly coupled ordinary differential equations (LCODEs), and food webs. Therefore, in Sect. 4.3, the output synchronization problem is analyzed for undirected and directed network models, respectively, with known coupling weights. Generally, complex networks with known coupling weights can not achieve synchronization by themselves; thus, it is necessary to design several control strategies to enable the network to achieve synchronization. Considering that the appropriate coupling weights can evidently improve the synchronizability of complex networks, two adaptive strategies are constructed to ensure the output synchronization of undirected and directed network models, respectively, by updating the coupling weights.

4.5 Numerical Examples

Example 4.13 A network model with five nodes is chosen with the following equation for each node:

$$\begin{pmatrix} \dot{x}_{i1} \\ \dot{x}_{i2} \\ \dot{x}_{i3} \end{pmatrix} = \begin{pmatrix} -x_{i1} - x_{i2} + \gamma(x_{i1}) \\ x_{i1} - 2x_{i2} + 2x_{i3} \\ -2x_{i3} \end{pmatrix}$$

where $\gamma(x_{i1}) = \frac{1}{4}(|x_{i1}+1| - |x_{i1}-1|) + 2$, $i = 1, 2, \cdots, 5$. Obviously, (A1) holds under the condition that $P = I_3$, $\Delta = \text{diag}(0, 0, 0)$ and $\eta = \frac{1}{2}$.

Let $c = 0.3$. Γ, C and $G(0)$, respectively, are as follows:

$$\Gamma = \begin{pmatrix} 0.7 & 0 & 0 \\ 0 & 0.6 & 0 \\ 0 & 0 & 0.8 \end{pmatrix},$$

$$C = \begin{pmatrix} 1 & 0 & 0 \\ 0 & 1 & 0 \end{pmatrix},$$

$$G(0) = \begin{pmatrix} -0.5 & 0.5 & 0 & 0 & 0 \\ 0 & -0.2 & 0 & 0.2 & 0 \\ 0 & 0 & -0.4 & 0 & 0.4 \\ 0 & 0.1 & 0.5 & -0.6 & 0 \\ 0.5 & 0 & 0.1 & 0 & -0.6 \end{pmatrix}.$$

Based on the MATLAB function NULL, $\theta = (1, 3, 1.5, 1, 1)^T \in \mathbb{R}^5$ can be found, which satisfies $\theta^T G(0) = 0$.

By simple calculations, the following can be obtained:

$$\hat{H} = \begin{pmatrix} 1 & 0 & 0 \\ 0 & 1 & 0 \end{pmatrix},$$

$$\xi = \text{diag}(\frac{2}{15}, \frac{2}{5}, \frac{1}{5}, \frac{2}{15}, \frac{2}{15}),$$

$$M = \begin{pmatrix} -\frac{2}{15} & \frac{1}{15} & 0 & 0 & \frac{1}{15} \\ \frac{1}{15} & -\frac{1}{15} & 0 & 0 & 0 \\ 0 & 0 & -\frac{1}{15} & \frac{1}{15} & 0 \\ 0 & 0 & \frac{1}{15} & -\frac{1}{15} & 0 \\ \frac{1}{15} & 0 & 0 & 0 & -\frac{1}{15} \end{pmatrix}.$$

Then, the following matrices can be found to satisfy (4.30):

$$Q = 7.6507 I_{10} > 0,$$

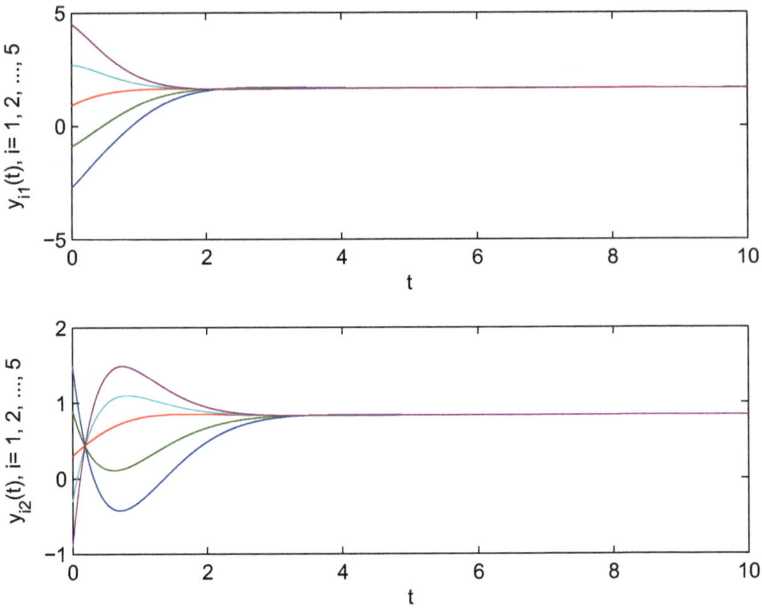

Fig. 4.1 $y_{i1}(t)$ and $y_{i2}(t)$, $i = 1, 2, \cdots, 5$, $t \in [0, 10s]$

$$\kappa = \begin{pmatrix} 0 & 0 & 0 & 0 & 0 \\ 0 & -2.9311 & 0 & 2.9311 & 0 \\ 0 & 0 & -3.5108 & 0 & 3.5108 \\ 0 & 2.9311 & 0 & -2.9311 & 0 \\ 0 & 0 & 3.5108 & 0 & -3.5108 \end{pmatrix}.$$

According to Theorem 4.9, based on the parameters given above, the complex network (4.22) achieves output synchronization by applying the adaptive law (4.24). Choose $\beta_{24} = \beta_{42} = 0.2$ and $\beta_{35} = \beta_{53} = 0.3$. Figs. 4.1 and 4.2 show the simulation results.

Example 4.14 A network model with five nodes is chosen with the following equation for each node (see [18]):

$$\begin{pmatrix} \dot{x}_{i1} \\ \dot{x}_{i2} \\ \dot{x}_{i3} \end{pmatrix} = \begin{pmatrix} 10(-x_{i1} + x_{i2} - g(x_{i1})) \\ x_{i1} - x_{i2} + x_{i3} \\ -14.87x_{i2} \end{pmatrix}$$

where $i = 1, 2, \cdots, 5$, $g(x_{i1}) = -0.68x_{i1} + 0.5(-1.27 + 0.68)(|x_{i1} + 1| - |x_{i1} - 1|)$. (A1) is clearly satisfied with $P = I_3$, $\eta = 1$ and $\Delta = \text{diag}(10, 13, 8)$.

Let $c = 0.3$. Γ, C and $G(0)$, respectively, are as follows:

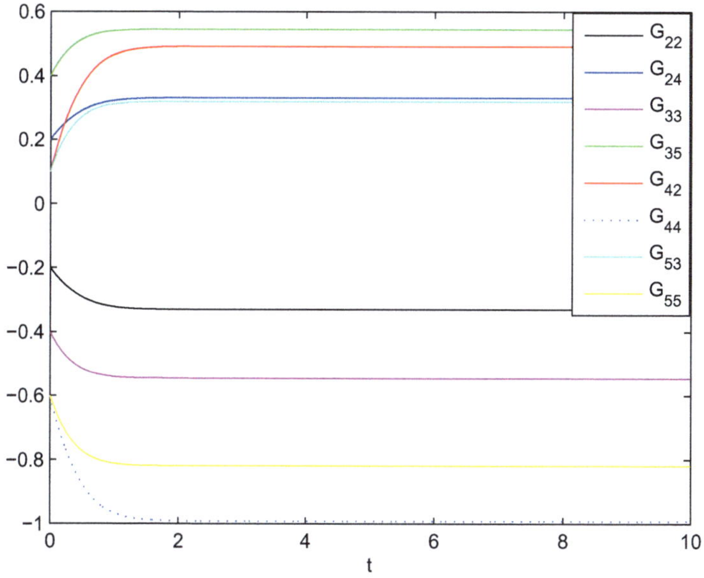

Fig. 4.2 Updated coupling weights $G_{ij}(t)$, $t \in [0, 10s]$

$$\Gamma = \begin{pmatrix} 0.5 & 0 & 0 \\ 0 & 0.3 & 0 \\ 0 & 0 & 0.4 \end{pmatrix},$$

$$C = \begin{pmatrix} 1 & 0 & 0 \\ 0 & 1 & 0 \end{pmatrix},$$

$$G(0) = \begin{pmatrix} -0.2 & 0.1 & 0 & 0.1 & 0 \\ 0.1 & -0.5 & 0 & 0 & 0.4 \\ 0 & 0 & -0.3 & 0.3 & 0 \\ 0.1 & 0 & 0.3 & -0.4 & 0 \\ 0 & 0.4 & 0 & 0 & -0.4 \end{pmatrix}.$$

Choose $\beta_{12} = \beta_{21} = 0.2$, $\beta_{14} = \beta_{41} = 0.4$, $\beta_{25} = \beta_{52} = 0.3$, and $\beta_{34} = \beta_{43} = 0.1$. According to Theorem 4.11, based on the parameters given above, the complex network (4.36) achieves output synchronization by applying the adaptive law (4.37). Figs. 4.3 and 4.4 show the simulation results.

Remark 4.15 Figs 4.1 and 4.3 present the output variables $y_{i1}(t)$ and $y_{i2}(t), i = 1, 2, \cdots, 5$ for CDNs (4.22) and (4.36), with $y_{i1}(t)$ and $y_{i2}(t), i = 1, 2, \cdots, 5$ shown to be synchronized. In Figs. 4.2 and 4.4, the curves of the adaptive coupling weights of CDNs (4.22) and (4.36), respectively, are plotted. As shown in the figures, each adaptive coupling weight gradually converges to a finite value.

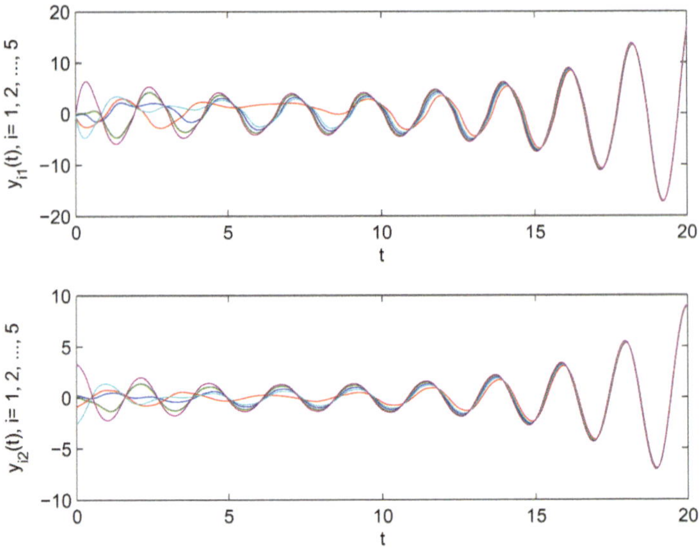

Fig. 4.3 $y_{i1}(t)$ and $y_{i2}(t)$, $i = 1, 2, \cdots, 5$, $t \in [0, 20s]$

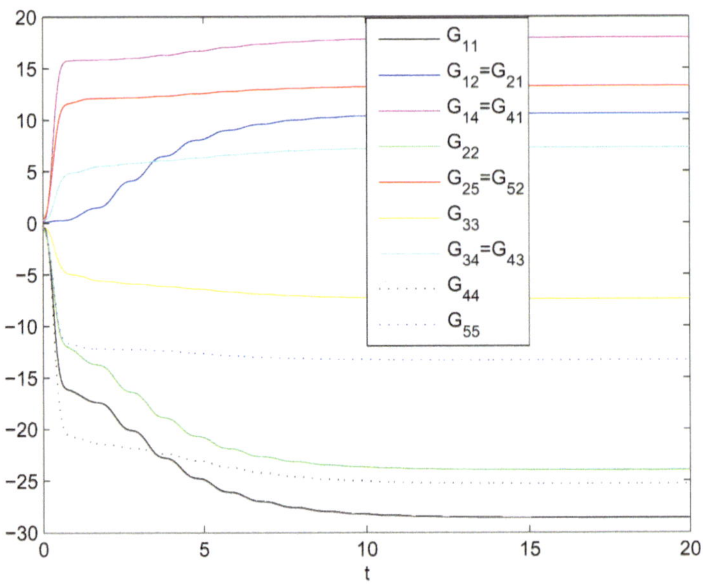

Fig. 4.4 Updated coupling weights $G_{ij}(t)$, $t \in [0, 20s]$

Remark 4.16 In [18], the authors analyzed the state synchronization of a network of Chua's circuits with given coupling coefficients (coupling weights and coupling strength). The output synchronization problem for an output coupled complex network model with fixed coupling weights was studied in [24]. In [18] and [24], the given coupling coefficients were larger than the values needed to achieve synchronization. Obviously, it is beneficial to make coupling coefficients small enough. In this chapter, the adaptive control method is used to achieve this goal, and two adaptive strategies are designed to obtain the appropriate coupling weights (see Figs. 4.2 and 4.4). It will provide the foundation for designing the appropriate coupling coefficients for output synchronization in CDNs, such as multi-agent systems and coupled neural networks.

4.6 Conclusion

This chapter studied the problem of output synchronization in complex network models with directed and undirected topologies. By applying Barbalat's lemma and constructing the appropriate Lyapunov functionals, several output synchronization criteria have been developed for these network models. Furthermore, some control schemes have been established to adjust the coupling weights among network nodes, and several criteria for the output synchronization have been deduced by applying the designed adaptive laws. The simulation results verify the effectiveness and correctness of the established output synchronization criteria. In future works, the output synchronization problem in delayed complex network models with directed and undirected topologies will be considered.

Chapter 5
Output Synchronization in CNNs With and Without External Disturbances

5.1 Introduction

In the past two decades, analysis and control for dynamical behaviors of various neural networks have been extensively discussed owing to their wide range of applications. In particular, a large number of researchers have taken into account the stability and stabilization problems of neural networks [16, 121–127]. Wang and Tian [122] proposed several criteria to guarantee global Lagrange stability of inertial neural networks by exploiting inequality method. In [124], Chen et al. discussed the mean-square global exponential stability for delayed recurrent neural networks with Markovian switching. Wei et al. [127] considered the impulsive stabilization of neural networks with discrete time-varying delays and unbounded continuously distributed delays.

More recently, coupled neural networks (CNNs) has been widely studied by researchers, which is consisting of several neural networks. The main reason for this is that CNNs has found fruitful applications in different fields such as chaos generators design, harmonic oscillation generation, secure communication, and so on. Particularly, synchronization problem of CNNs has received much of the focus in recent years [91, 128–141]. Feng et al. [128] discussed stochastic synchronization for CNNs with hybrid nonlinear coupling by exploiting the Kronecker product technique and Lyapunov functional method. In [134], the authors studied global exponential synchronization problem of delayed CNNs by employing matrix inequality technique and Lyapunov functional. Yuan [137] considered synchronization of an uncertain CNNs with distributed-delay coupling and constant coupling. In [138], a Markovian CNNs with random coupling strengths was discussed. Zhang et al. [139] derived several delay-dependent synchronization criteria for Markovian hybrid CNNs with time-varying mode-delays. In [91], the authors investigated the finite-time synchronization problem of nonlinear CNNs. By employing adaptive feedback control method and Lyapunov stability theorem, some sufficient conditions for ensuring the exponential synchronization of Markovian

J.-L. Wang et al., *Analysis and Control of Output Synchronization for Complex Dynamical Networks*, https://doi.org/10.1007/978-981-13-1352-3_5

stochastic coupled neural networks of neutral-type were presented in [140]. Wang et al. [141] obtained several sampled-data synchronization criteria for CNNs by applying Wirtinger's inequality and Jensen's integral inequality. It is worth noting that the state synchronization of CNNs was considered in these papers [91, 128–141]. As is well known, state synchronization of CNNs is of course a very preferable property but is a very difficult problem. Furthermore, in many situations, only partial states are needed to make the synchronization to come true. Obviously, it is more meaningful to investigate output synchronization of CNNs. To the best of our knowledge, very few results on this topic have been reported.

On the other hand, external disturbances widely exist in complex network models, which may destroy the network synchronization. Hence, many authors have investigated the $\mathcal{H}_\infty$ synchronization problem for complex networks [142–146]. In [142], the authors considered $\mathcal{H}_\infty$ synchronization problem for delayed CDNs by employing sampled-data controllers. Shen et al. [143] discussed the finite-time $\mathcal{H}_\infty$ synchronization problem of delayed CDNs with semi-Markov jump topology. Nevertheless, there are very few researches about the $\mathcal{H}_\infty$ output synchronization for complex networks. Consequently, it is also very significant to investigate $\mathcal{H}_\infty$ output synchronization problem of CNNs.

As is known to us, coupling weights have an important influence on network synchronization, and complex networks can realize synchronization in many situations as long as the coupling weights are large enough [88]. Generally speaking, complex dynamic network is not synchronized with given coupling weights. For these phenomena, some researchers presented several adaptive schemes for synchronization of CDNs by adjusting the coupling weights [11, 108, 109, 147, 148]. In [109], the authors studied the synchronization of networks of nonlinear oscillators by exploiting edge-based adaptive strategies. Yu et al. [11] respectively presented two adaptive schemes for updating partial or all coupling weights such that complex networks can achieve synchronization. In [148], Xu et al. studied the synchronization problem for fractional-order complex networks with adaptive coupling by exploiting Laplace transform and Mittag-Leffler function. Obviously, it is of great importance to take into account synchronization of CNNs by tuning the coupling weights [51]. Wang et al. [51] considered the passivity of a CNNs by designing suitable adaptive laws. By utilizing the relationship between synchronization and passivity, a global synchronization criterion was derived. Unfortunately, in these existing works [11, 17, 51, 108, 109, 147–149], the state synchronization of CDNs was studied. As far as we know, very few authors have discussed output synchronization of CNNs with adaptive coupling weights. Especially, $\mathcal{H}_\infty$ output synchronization for CNNs with adaptive coupling weights has not yet been taken into account.

In this chapter, we respectively study output synchronization and $\mathcal{H}_\infty$ output synchronization of CNNs with directed and undirected topology. The main contributions of this chapter are as follows. First, we analyze output synchronization of CNNs by employing matrix theory and Barbalat's Lemma. Second, two adaptive strategies are proposed to guarantee the output synchronization of CNNs. Third, we discuss the $\mathcal{H}_\infty$ output synchronization of CNNs with external disturbances by employing Lyapunov functionals and matrix theory. Fourth, several adaptive strategies for $\mathcal{H}_\infty$ output synchronization of CNNs are also designed.

5.2 Output Synchronization of CNNs with Fixed and Adaptive Couplings

5.2.1 Network Model

The network model considered in this section is described by:

$$\dot{z}_i(t) = -Az_i(t) + Bg(z_i(t)) + J + c\sum_{j=1}^{N} G_{ij}\Gamma z_j(t),$$
$$y_i(t) = Hz_i(t), \quad i = 1, 2, \cdots, N, \tag{5.1}$$

where $z_i(t) = (z_{i1}(t), z_{i2}(t), \cdots, z_{in}(t))^T \in \mathbb{R}^n$ represents the state vector of the ith node, $y_i(t) = (y_{i1}(t), y_{i2}(t), \cdots, y_{iq}(t))^T \in \mathbb{R}^q$ is the output vector of the ith node, $A = \text{diag}(a_1, a_2, \cdots, a_n) \in \mathbb{R}^{n\times n} > 0$, $B \in \mathbb{R}^{n\times n}$ and $H \in \mathbb{R}^{q\times n}$ are constant matrices, $g(z_i(t)) = (g_1(z_{i1}(t)), g_2(z_{i2}(t)), \cdots, g_n(z_{in}(t)))^T \in \mathbb{R}^n$ and $J = (J_1, J_2, \cdots, J_n)^T \in \mathbb{R}^n$, $\mathbb{R} \ni c > 0$ is overall coupling strength, $\Gamma \in \mathbb{R}^{n\times n} > 0$ denotes the inner coupling structure, $G = (G_{ij})_{N\times N}$ represents the coupling weights: if there is a connection from node i to node $j (i \neq j)$, then $G_{ij} > 0$, otherwise, $G_{ij} = 0 \ (i \neq j)$ and

$$G_{ii} = -\sum_{\substack{j=1\\ j\neq i}}^{N} G_{ij}.$$

In this chapter, CNNs (5.1) is strongly connected and function $g_j(\cdot)$ satisfies

$$|g_j(\gamma_1) - g_j(\gamma_2)| \leqslant \beta_j|\gamma_1 - \gamma_2|$$

for any $\gamma_1, \gamma_2 \in \mathbb{R}$, where β_j is a positive constant, $j = 1, 2, \cdots, n$. Take $\Delta = \text{diag}(\beta_1^2, \beta_2^2, \cdots, \beta_n^2)$.

Remark 5.1 In this section, a CNNs model is discussed. Recently, the state synchronization problem of CNNs has been studied by many researchers. Practically, state in CNNs is very hard to be observed in many circumstances. Furthermore, in many situations, only partial states are needed to make the synchronization to come true. Obviously, it is more interesting and important to further investigate output synchronization of CNNs.

Definition 5.2 (see [24]) The outputs of CNNs (5.1) are synchronized if

$$\lim_{t \to +\infty} \|y_i(t) - y_j(t)\| = 0 \text{ for all } i, j = 1, 2, ..., N.$$

5.2.2 *Output Synchronization of CNNs with Fixed Coupling*

5.2.2.1 Directed Topology

By Lemma 1.10, we can easily find a positive vector $(\varrho_1, \varrho_2, \cdots, \varrho_N)^T \in \mathbb{R}^N$ such that

$$(\varrho_1, \varrho_2, \cdots, \varrho_N)G = 0.$$

Define $\lambda_i = \frac{\varrho_i}{\varrho_1+\varrho_2+\cdots+\varrho_N}$ and $z^*(t) = \sum_{i=1}^N \lambda_i z_i(t)$. Then, one obtains

$$\begin{aligned}
\dot{z}^*(t) =& -A\sum_{i=1}^N \lambda_i z_i(t) + B\sum_{i=1}^N \lambda_i g(z_i(t)) + \sum_{i=1}^N \lambda_i J \\
& + c\sum_{j=1}^N \left(\sum_{i=1}^N \lambda_i G_{ij}\right) \Gamma z_j(t) \\
=& -Az^*(t) + B\sum_{i=1}^N \lambda_i g(z_i(t)) + J. \qquad (5.2)
\end{aligned}$$

Let $e_i(t) = z_i(t) - z^*(t)$. By (5.1) and (5.2), we can get

$$\begin{aligned}
\dot{e}_i(t) =& -Ae_i(t) + Bg(z_i(t)) - B\sum_{k=1}^N \lambda_k g(z_k(t)) \\
& + c\sum_{j=1}^N G_{ij}\Gamma e_j(t), \qquad (5.3)
\end{aligned}$$

where $i = 1, 2, \cdots, N$.

Furthermore, we can easily find the nonsingular matrices $H_1 \in \mathbb{R}^{q\times q}$ and $H_2 \in \mathbb{R}^{n\times n}$ satisfying

$$H = H_1 \begin{pmatrix} I_r & 0 \\ 0 & 0 \end{pmatrix} H_2,$$

where $r = \text{Rank}(H)$. Then, one has

$$\dot{e}_i(t) = -Ae_i(t) + Bg(z_i(t)) - B\sum_{k=1}^{N}\lambda_k g(z_k(t)) + c\sum_{j=1}^{N}G_{ij}\Gamma e_j(t),$$

$$y_i(t) = H_1\begin{pmatrix} I_r & 0 \\ 0 & 0 \end{pmatrix}H_2 e_i(t) + Hz^*(t), \tag{5.4}$$

where $i = 1, 2, \cdots, N$.

Therefore, the outputs of CNNs (5.1) are synchronized if and only if $\hat{y}_i(t)$ in (5.5) asymptotically converge to zero.

$$\begin{aligned}\dot{e}_i(t) &= -Ae_i(t) + Bg(z_i(t)) - B\sum_{k=1}^{N}\lambda_k g(z_k(t)) \\ &\quad + c\sum_{j=1}^{N}G_{ij}\Gamma e_j(t), \\ \hat{y}_i(t) &= H_3 e_i(t),\end{aligned} \tag{5.5}$$

where $\hat{y}_i(t) = (\hat{y}_{i1}(t), \hat{y}_{i2}(t), \cdots, \hat{y}_{ir}(t))^T \in \mathbb{R}^r$, $H_3 = (I_r \ \ 0)_{r\times n}H_2$, $i = 1, 2, \cdots, N$.

Theorem 5.3 *The outputs of CNNs* (5.1) *are synchronized if there exist a matrix* $P \in \mathbb{R}^{n\times n} > 0$ *and a positive real number* κ *such that*

$$\lambda \otimes \left(-PA - AP + PBB^T P + \Delta + \kappa H_3^T H_3\right) + \mathbb{Q} \leqslant 0, \tag{5.6}$$

where $\lambda = \text{diag}(\lambda_1, \lambda_2, \cdots, \lambda_N)$, $\mathbb{Q} = (c\lambda G) \otimes (P\Gamma) + (cG^T\lambda) \otimes (\Gamma P)$.

Proof Choose the following Lyapunov functional for network (5.5):

$$V_1(e(t)) = \sum_{i=1}^{N}\lambda_i e_i^T(t)Pe_i(t),$$

where $e(t) = (e_1^T(t), e_2^T(t), \cdots, e_N^T(t))^T$. Then, one has

$$\begin{aligned}\dot{V}_1(e(t)) = 2\sum_{i=1}^{N}\lambda_i e_i^T(t)P[&-Ae_i(t) + Bg(z_i(t)) \\ &-Bg(z^*(t)) + Bg(z^*(t)) - B\sum_{k=1}^{N}\lambda_k g(z_k(t))\end{aligned}$$

$$+c\sum_{j=1}^{N} G_{ij}\Gamma e_j(t)]. \tag{5.7}$$

Obviously,

$$2\sum_{i=1}^{N}\lambda_i e_i^T(t)PB\left(g(z_i(t)) - g(z^*(t))\right)$$
$$\leqslant \sum_{i=1}^{N}\lambda_i e_i^T(t)\left(PBB^TP + \Delta\right)e_i(t). \tag{5.8}$$

On the other hand,

$$\begin{aligned}\sum_{i=1}^{N}\lambda_i e_i(t) &= \sum_{i=1}^{N}\lambda_i\left(z_i(t) - \sum_{j=1}^{N}\lambda_j z_j(t)\right)\\ &= \sum_{i=1}^{N}\lambda_i z_i(t) - \sum_{j=1}^{N}\left(\sum_{i=1}^{N}\lambda_i\right)\lambda_j z_j(t)\\ &= 0.\end{aligned} \tag{5.9}$$

By (5.9), one gets

$$\sum_{i=1}^{N}\lambda_i e_i^T(t)PB\left(g(z^*(t)) - \sum_{k=1}^{N}\lambda_k g(z_k(t))\right) = 0. \tag{5.10}$$

From (5.6), (5.7), (5.8) and (5.10), we can obtain

$$\begin{aligned}\dot{V}_1(e(t)) &\leqslant e^T(t)\Big[\lambda\otimes(-PA - AP + PBB^TP + \Delta)\\ &\quad +(c\lambda G)\otimes(P\Gamma) + (cG^T\lambda)\otimes(\Gamma P)\Big]e(t)\\ &\leqslant -\kappa\sum_{i=1}^{N}\lambda_i e_i^T(t)H_3^T H_3 e_i(t).\end{aligned} \tag{5.11}$$

From (5.11) and the definition of $V_1(e(t))$, we can conclude that

$$\lim_{t\to+\infty} V_1(e(t)) = \xi \in \mathbb{R} \geqslant 0, \tag{5.12}$$

$$\|e_i(t)\| \leqslant \alpha \in \mathbb{R} > 0,\ i = 1, 2, \cdots, N. \tag{5.13}$$

Take

$$\delta(t)=\sum_{i=1}^{N}\lambda_i e_i^T(t)H_3^T H_3 e_i(t).$$

By (5.11) and (5.12), one obtains

$$\begin{aligned}
&\lim_{t\to+\infty}\int_0^t \delta(s)ds \\
\leqslant &-\frac{1}{\kappa}\lim_{t\to+\infty}\int_0^t \dot{V}_1(e(s))ds \\
=&\frac{1}{\kappa}\Big(V_1(e(0))-\xi\Big).
\end{aligned} \tag{5.14}$$

Furthermore,

$$\begin{aligned}
\dot{\delta}(t)=&2\sum_{i=1}^{N}\lambda_i e_i^T(t)H_3^T H_3\Big[-Ae_i(t)+Bg(z_i(t)) \\
&-Bg(z^*(t))+Bg(z^*(t))-B\sum_{k=1}^{N}\lambda_k g(z_k(t)) \\
&+c\sum_{j=1}^{N}G_{ij}\Gamma e_j(t)\Big] \\
=&2\sum_{i=1}^{N}\lambda_i e_i^T(t)H_3^T H_3\Big[-Ae_i(t)+Bg(z_i(t)) \\
&-Bg(z^*(t))+c\sum_{j=1}^{N}G_{ij}\Gamma e_j(t)\Big].
\end{aligned} \tag{5.15}$$

From (5.13) and (5.15), one has

$$\begin{aligned}
\left|\dot{\delta}(t)\right|\leqslant &2\sum_{i=1}^{N}\lambda_i\left|e_i^T(t)H_3^T H_3 A e_i(t)\right| \\
&+2\sum_{i=1}^{N}\lambda_i\left|e_i^T(t)H_3^T H_3 B\left(g(z_i(t))-g(z^*(t))\right)\right|
\end{aligned}$$

$$+2c\sum_{i=1}^{N}\sum_{j=1}^{N}\lambda_i\left|G_{ij}\right|\left|e_i^T(t)H_3^T H_3\Gamma e_j(t)\right|$$

$$\leqslant c\sum_{i=1}^{N}\sum_{j=1}^{N}\left|G_{ij}\right|\left(\lambda_M^2(H_3^T H_3)+\lambda_M^2(\Gamma)\right)\alpha^2$$

$$+N\big(2\lambda_M^2(H_3^T H_3)+\lambda_M^2(A)$$

$$+\lambda_M(B^T B)\lambda_M(\Delta)\big)\alpha^2. \tag{5.16}$$

From (5.16), we can conclude that $\delta(t)$ is uniformly continuous. In view of Lemma 1.8, one has

$$\lim_{t\to+\infty}\delta(t)=0. \tag{5.17}$$

By (5.17), one obtains

$$\begin{aligned}&\lim_{t\to+\infty}\|\hat{y}_i(t)\|\\=&\lim_{t\to+\infty}\|H_3 e_i(t)\|\\=&\ 0,\ i=1,2,\cdots,N.\end{aligned} \tag{5.18}$$

Therefore, the outputs of CNNs (5.1) are synchronized. □

5.2.2.2 Undirected Topology

When the topological structure of CNNs (5.1) is undirected, G becomes a symmetric matrix. Then, one has

$$(1,1,\cdots,1)G=0.$$

Define $z^*(t)=\frac{1}{N}\sum_{i=1}^{N}z_i(t)$ and $e_i(t)=z_i(t)-z^*(t)$. Then, one obtains

$$\dot{e}_i(t)=-Ae_i(t)+Bg(z_i(t))-B\sum_{k=1}^{N}\frac{g(z_k(t))}{N}$$

$$+c\sum_{j=1}^{N}G_{ij}\Gamma e_j(t),\ i=1,2,\cdots,N. \tag{5.19}$$

Similarly, we can conclude that the outputs of CNNs (5.1) are synchronized if and only if $\hat{y}_i(t)$ in (5.20) asymptotically converge to zero.

$$\begin{aligned}
\dot{e}_i(t) &= -Ae_i(t) + Bg(z_i(t)) - B\sum_{k=1}^{N}\frac{g(z_k(t))}{N} \\
&\quad +c\sum_{j=1}^{N}G_{ij}\Gamma e_j(t), \\
\hat{y}_i(t) &= H_3 e_i(t),\ i = 1, 2, \cdots, N,
\end{aligned} \tag{5.20}$$

where H_3 and $\hat{y}_i(t)$ have the same meanings as defined in (5.5).

Theorem 5.4 *The outputs of CNNs* (5.1) *are synchronized if there exist a positive real number κ and a matrix $P \in \mathbb{R}^{n\times n} > 0$ such that*

$$I_N \otimes \left(-PA - AP + PBB^T P + \Delta + \kappa H_3^T H_3\right) + \hat{\mathbb{Q}} \leqslant 0,$$

where $\hat{\mathbb{Q}} = cG \otimes (P\Gamma + \Gamma P)$.

Proof Choose the following Lyapunov functional for network (5.20):

$$V_2(e(t)) = \sum_{i=1}^{N} e_i^T(t)Pe_i(t),$$

where $e(t) = (e_1^T(t), e_2^T(t), \cdots, e_N^T(t))^T$. Then, one has

$$\begin{aligned}
\dot{V}_2(e(t)) &\leqslant \sum_{i=1}^{N} e_i^T(t)(-PA - AP + PBB^T P + \Delta)e_i(t) \\
&\quad +2c\sum_{i=1}^{N}\sum_{j=1}^{N} G_{ij}e_i^T(t)P\Gamma e_j(t) \\
&= e^T(t)\big[I_N \otimes (-PA - AP + PBB^T P + \Delta) \\
&\quad +cG \otimes (P\Gamma + \Gamma P)\big]e(t) \\
&\leqslant -\kappa\sum_{i=1}^{N} e_i^T(t)H_3^T H_3 e_i(t).
\end{aligned}$$

Similarly, we can conclude that the outputs of CNNs (5.1) are synchronized. □

Remark 5.5 Based on a special solution of an isolated node, some researchers discussed output synchronization of CDNs with fixed coupling [24]. In this subsection, we remove this restriction and analyze the output synchronization of CNNs with fixed coupling. By choosing appropriate Lyapunov functionals and employing Barbalat's lemma, we establish two sufficient conditions to guarantee that the outputs of CNNs (5.1) are synchronized in the scenarios with directed and undirected topologies (see Theorems 5.3 and 5.4).

5.2.3 Output Synchronization of CNNs with Adaptive Coupling

5.2.3.1 Directed Topology

By Lemma 1.10, there obviously exists a positive vector $(\varrho_1, \varrho_2, \cdots, \varrho_N)^T \in \mathbb{R}^N$ such that

$$(\varrho_1, \varrho_2, \cdots, \varrho_N)G(0) = 0.$$

Take

$$\lambda_i = \frac{\varrho_i}{\varrho_1 + \varrho_2 + \cdots + \varrho_N}.$$

Then, one gets

$$\sum_{j=1}^{N} U_{ij} = \sum_{j=1}^{N} U_{ji} = 0,$$

where $\lambda = \text{diag}(\lambda_1, \lambda_2, \cdots, \lambda_N)$, $U = (U_{ij})_{N\times N} = \lambda G(0) + G^T(0)\lambda$.

In order to guarantee the output synchronization of CNNs (5.1), the following adaptive strategy for updating the coupling weights is designed:

$$\dot{G}_{ij}(t) = \begin{cases} \varpi_{ij}\lambda_j(z_i(t) - z_j(t))^T \Gamma(z_i(t) - z_j(t)), \\ \qquad\qquad \text{if } (i, j) \in \mathcal{E} \text{ and } (j, i) \in \mathcal{E}, \\ -\sum_{\substack{k=1 \\ k\neq i}}^{N} \dot{G}_{ik}(t), \qquad \text{if } i = j, \\ 0, \qquad\qquad \text{otherwise,} \end{cases} \tag{5.21}$$

where $\varpi_{ij} = \varpi_{ji}$ are positive real numbers, $\mathcal{E} \subset \{1, 2, \cdots, N\} \times \{1, 2, \cdots, N\}$ denotes the set of directed edges in CNNs (5.1).

From (5.21), one gets

$$\sum_{j=1}^{N} \dot{G}_{ij}(t) = 0, \ i = 1, 2, \cdots, N,$$

$$\sum_{j=1}^{N} \lambda_j \dot{G}_{ji}(t) = \sum_{j=1}^{N} \lambda_i \dot{G}_{ij}(t) = 0, \ i = 1, 2, \cdots, N,$$

for any $t \in [0, +\infty)$.

Since $(\lambda_1, \lambda_2, \cdots, \lambda_N)G(0) = 0$, $G_{ii}(0) = -\sum_{\substack{j=1\\ j\neq i}}^{N} G_{ij}(0)$ and $\sum_{j=1}^{N} U_{ij} = 0$, one has

$$\sum_{j=1}^{N} \lambda_j G_{ji}(t) = 0, \ i = 1, 2, \cdots, N, \tag{5.22}$$

$$G_{ii}(t) = -\sum_{\substack{j=1\\ j\neq i}}^{N} G_{ij}(t), \ i = 1, 2, \cdots, N, \tag{5.23}$$

$$\sum_{j=1}^{N} \Big(\lambda_i G_{ij}(t) + \lambda_j G_{ji}(t)\Big) = 0, \tag{5.24}$$

for any $t \in [0, +\infty)$.

Define $z^*(t) = \sum_{i=1}^{N} \lambda_i z_i(t)$. From (5.1) and (5.22), one obtains

$$\begin{aligned}
\dot{z}^*(t) =& -A\sum_{i=1}^{N} \lambda_i z_i(t) + B\sum_{i=1}^{N} \lambda_i g(z_i(t)) + \sum_{i=1}^{N} \lambda_i J \\
&+ c\sum_{j=1}^{N} \left(\sum_{i=1}^{N} \lambda_i G_{ij}(t)\right) \Gamma z_j(t) \\
=& -Az^*(t) + B\sum_{i=1}^{N} \lambda_i g(z_i(t)) + J.
\end{aligned} \tag{5.25}$$

Let $e_i(t) = z_i(t) - z^*(t)$. By (5.1), (5.23) and (5.25), we can get

$$\begin{aligned}
\dot{e}_i(t) =& -Ae_i(t) + Bg(z_i(t)) - B\sum_{k=1}^{N} \lambda_k g(z_k(t)) \\
&+ c\sum_{j=1}^{N} G_{ij}(t)\Gamma e_j(t), \ i = 1, 2, \cdots, N.
\end{aligned}$$

Similarly, we can conclude that the outputs of CNNs (5.1) are synchronized under the adaptive strategy (5.21) if and only if $\hat{y}_i(t)$ in (5.26) asymptotically converge to zero.

$$\dot{e}_i(t) = -Ae_i(t) + Bg(z_i(t)) - B\sum_{k=1}^{N} \lambda_k g(z_k(t))$$

$$+c\sum_{j=1}^{N} G_{ij}\Gamma e_j(t),$$

$$\hat{y}_i(t) = H_3 e_i(t),\ i = 1, 2, \cdots, N, \tag{5.26}$$

where H_3 and $\hat{y}_i(t)$ have the same meanings as in (5.5).

Denote

$$F_{ij} = \begin{cases} \lambda_i G_{ij}(0), & \text{if } (i,j) \in \mathcal{E} \text{ and } (j,i) \notin \mathcal{E}, \\ \lambda_j G_{ji}(0), & \text{if } (j,i) \in \mathcal{E} \text{ and } (i,j) \notin \mathcal{E}, \\ -\sum_{\substack{k=1 \\ k\neq i}}^{N} F_{ik}, & \text{if } i = j, \\ 0, & \text{otherwise.} \end{cases}$$

Theorem 5.6 *The outputs of CNNs* (5.1) *are synchronized under the adaptive strategy* (5.21) *if there exist a positive real number* κ *and a matrix* $M = (M_{ij})_{N\times N} \in \mathbb{R}^{N\times N}$ *such that*

$$\lambda \otimes (-2A + BB^T + \Delta + \kappa H_3^T H_3) + c(F + M) \otimes \Gamma \leqslant 0,$$

where $F = (F_{ij})_{N\times N} \in \mathbb{R}^{N\times N}$, $M_{ij} = M_{ji} (i \neq j)$ *are nonnegative real numbers,* $M_{ii} = -\sum_{\substack{j=1 \\ j\neq i}}^{N} M_{ij}$, *and* $M_{ij} = 0 (i \neq j)$ *if and only if* $G_{ij}(0) \times G_{ji}(0) = 0$.

Proof Choose the following Lyapunov functional for system (5.26):

$$V_3(e(t)) = \sum_{i=1}^{N} \lambda_i e_i^T(t) e_i(t) + c\sum_{i=1}^{N} \sum_{\substack{(i,j)\in\mathcal{E}, \\ (j,i)\in\mathcal{E}}} \frac{(\lambda_i G_{ij}(t) + \lambda_j G_{ji}(t) - M_{ij})^2}{8\varpi_{ij}\lambda_i\lambda_j},$$

where $e(t) = (e_1^T(t), e_2^T(t), \cdots, e_N^T(t))^T$.

Then, one has

$$\dot{V}_3(e(t)) \leqslant \sum_{i=1}^{N} \lambda_i e_i^T(t)(-2A + BB^T + \Delta) e_i(t) + c\sum_{i=1}^{N}\sum_{j=1}^{N} (\lambda_i G_{ij}(t) + \lambda_j G_{ji}(t)) e_i^T(t)\Gamma e_j(t)$$

$$
\begin{aligned}
&+\frac{c}{2}\sum_{i=1}^{N}\sum_{\substack{(i,j)\in\mathcal{E},\\(j,i)\in\mathcal{E}}}(\lambda_i G_{ij}(t)+\lambda_j G_{ji}(t)-M_{ij})(e_i(t)\\
&-e_j(t))^T\Gamma(e_i(t)-e_j(t))\\
&=e^T(t)\Big[\lambda\otimes(-2A+BB^T+\Delta)\\
&+c(F+M)\otimes\Gamma\Big]e(t)\\
&\leqslant -\kappa\sum_{i=1}^{N}\lambda_i e_i^T(t)H_3^T H_3 e_i(t).
\end{aligned}
\tag{5.27}
$$

From (5.21), (5.27) and the definition of $V_3(e(t))$, we can get

$$
\left|G_{ij}(t)\right|\leqslant\zeta\in\mathbb{R}>0,\ i,j=1,2,\cdots,N,
$$

for any $t\in[0,+\infty)$.

Similarly, we can conclude that the outputs of CNNs (5.1) are synchronized under the adaptive strategy (5.21). □

5.2.3.2 Undirected Topology

When the CNNs (5.1) is undirected, $G(0)$ becomes a symmetric matrix. Then, one has

$$
(1,1,\cdots,1)G(0)=0.
$$

In order to guarantee that the outputs of CNNs (5.1) are synchronized, the following adaptive strategy for updating the coupling weights is designed:

$$
\dot{G}_{ij}(t)=\begin{cases}
\varpi_{ij}(z_i(t)-z_j(t))^T\Gamma(z_i(t)-z_j(t)), & \text{if } (i,j)\in\mathcal{E},\\
-\sum_{\substack{k=1\\k\neq i}}^{N}\dot{G}_{ik}(t), & \text{if } i=j,\\
0, & \text{otherwise},
\end{cases}
\tag{5.28}
$$

where $\varpi_{ij}=\varpi_{ji}$ are positive real numbers, $\mathcal{E}$ denotes the set of undirected edges in CNNs (5.1).

Define $z^*(t)=\frac{1}{N}\sum_{i=1}^{N}z_i(t)$ and $e_i(t)=z_i(t)-z^*(t)$. Then, we can obtain

$$
\dot{e}_i(t)=-Ae_i(t)+Bg(z_i(t))-B\sum_{k=1}^{N}\frac{g(z_k(t))}{N}
$$

$$+c\sum_{j=1}^{N} G_{ij}(t)\Gamma e_j(t),\ i = 1, 2, \cdots, N. \tag{5.29}$$

Similarly, we can conclude that the outputs of CNNs (5.1) are synchronized under the adaptive strategy (5.28) if and only if $\hat{y}_i(t)$ in (5.30) asymptotically converge to zero.

$$\begin{aligned}\dot{e}_i(t) &= -Ae_i(t) + Bg(z_i(t)) - B\sum_{k=1}^{N}\frac{g(z_k(t))}{N} \\ &\quad +c\sum_{j=1}^{N} G_{ij}(t)\Gamma e_j(t), \\ \hat{y}_i(t) &= H_3 e_i(t),\end{aligned} \tag{5.30}$$

where $i = 1, 2, \cdots, N$, H_3 and $\hat{y}_i(t)$ have the same meanings as in (5.5).

Theorem 5.7 *The outputs of CNNs* (5.1) *are synchronized under the adaptive strategy* (5.28).

Proof Choose the following Lyapunov functional for system (5.30):

$$\begin{aligned}V_4(e(t)) &= \sum_{i=1}^{N} e_i^T(t)e_i(t) \\ &\quad +c\sum_{i=1}^{N}\sum_{(i,j)\in\mathcal{E}}\frac{(G_{ij}(t) - M_{ij})^2}{2\varpi_{ij}},\end{aligned}$$

where $e(t) = (e_1^T(t), e_2^T(t), \cdots, e_N^T(t))^T$, $M_{ij} = M_{ji}(i \neq j)$ are nonnegative real numbers, $M_{ii} = -\sum_{\substack{j=1\\ j\neq i}}^{N} M_{ij}$, and $M_{ij} = 0(i \neq j)$ if and only if $G_{ij}(0) = 0$.

Then, one obtains

$$\begin{aligned}\dot{V}_4(e(t)) &\leqslant e^T(t)\Big[I_N \otimes (-2A + BB^T + \Delta) \\ &\quad +2cM \otimes \Gamma\Big]e(t),\end{aligned}$$

where $M = (M_{ij})_{N\times N} \in \mathbb{R}^{N\times N}$.

On the other hand, we can find a unitary matrix $\Theta = (\theta_1, \theta_2, \cdots, \theta_N) \in \mathbb{R}^{N\times N}$ satisfying

$$\Theta^T M\Theta = \mu = \text{diag}(\mu_1, \mu_2, \cdots, \mu_N),$$

where $0 = \mu_1 > \mu_2 \geqslant \mu_3 \geqslant \cdots \geqslant \mu_N$.

Define

$$\upsilon(t) = (\upsilon_1^T(t), \upsilon_2^T(t), \cdots, \upsilon_N^T(t))^T = (\Theta^T \otimes I_n)e(t).$$

Since $\theta_1 = \frac{1}{\sqrt{N}}(1, 1, \cdots, 1)^T$, we can easily derive $\upsilon_1(t) = (\theta_1^T \otimes I_n)e(t) = 0$. Then, we have

$$\begin{aligned}
\dot{V}_4(e(t)) &\leqslant e^T(t)\Big[I_N \otimes (-2A + BB^T + \Delta) \\
&\quad +2c(\Theta \otimes I_n)(\mu \otimes \Gamma)(\Theta^T \otimes I_n)\Big]e(t) \\
&= e^T(t)[I_N \otimes (-2A + BB^T + \Delta)]e(t) \\
&\quad +2c\upsilon^T(t)(\mu \otimes \Gamma)\upsilon(t) \\
&\leqslant e^T(t)[I_N \otimes (-2A + BB^T + \Delta)]e(t) \\
&\quad +2c\mu_2\upsilon^T(t)(I_N \otimes \Gamma)\upsilon(t) \\
&= e^T(t)[I_N \otimes (-2A + BB^T + \Delta + 2c\mu_2\Gamma)]e(t).
\end{aligned}$$

By selecting M_{ij} large enough such that

$$-2A + BB^T + \Delta + H_3^T H_3 + 2c\mu_2\Gamma \leqslant 0,$$

one has

$$\dot{V}_4(e(t)) \leqslant -\sum_{i=1}^{N} e_i^T(t)H_3^T H_3 e_i(t).$$

Similarly, we can conclude that the outputs of CNNs (5.1) are synchronized under the adaptive strategy (5.28). □

Remark 5.8 In subsection 5.2.2, we respectively analyze the output synchronization of directed and undirected CNNs with fixed coupling. However, the outputs of CNNs with given coupling weights are not synchronized in many circumstances, thus two adaptive strategies for updating the coupling weights are developed in this subsection to guarantee the output synchronization of CNNs (see (5.21) and (5.28)). By employing the designed adaptive strategies (5.21), (5.28) and the Barbalat's Lemma, we establish two sufficient conditions to ensure that the outputs of CNNs (5.1) are synchronized in the scenarios with directed and undirected topologies (see Theorems 5.6 and 5.7).

5.3 $\mathcal{H}_\infty$ Output Synchronization of CNNs with Fixed and Adaptive Couplings

5.3.1 Network Model

The network model considered in this section is described by:

$$\begin{aligned}\dot{z}_i(t) &= -Az_i(t) + Bg(z_i(t)) + J + w_i(t) \\ &\quad +c\sum_{j=1}^{N} G_{ij}\Gamma z_j(t), \\ y_i(t) &= Hz_i(t), \ \ i = 1, 2, \cdots, N,\end{aligned} \tag{5.31}$$

where $w_i(t) \in \mathbb{R}^n$ denotes the external disturbance and is square integrable, i.e.,

$$\int_0^{t_p} w_i^T(t)w_i(t)dt < +\infty$$

for any $\mathbb{R} \ni t_p \geqslant 0$. Take $w(t) = (w_1^T(t), w_2^T(t), \cdots, w_N^T(t))^T$.

Remark 5.9 As far as we know, the external disturbance is ubiquitous, and may destroy the output synchronization of CNNs, thus it is essential to consider the external disturbance in CNNs (5.1). In this section, the $\mathcal{H}_\infty$ output synchronization of CNNs (5.1) with external disturbances $w_i(t)$, $i = 1, 2, \cdots, N$, is discussed (see [108]).

5.3.2 $\mathcal{H}_\infty$ Output Synchronization of CNNs with Fixed Coupling

5.3.2.1 Directed Topology

In view of Lemma 1.10, we can find a positive vector $(\varrho_1, \varrho_2, \cdots, \varrho_N)^T \in \mathbb{R}^N$ satisfying

$$(\varrho_1, \varrho_2, \cdots, \varrho_N)G = 0.$$

Let $\lambda_i = \frac{\varrho_i}{\varrho_1+\varrho_2+\cdots+\varrho_N}$, $z^*(t) = \sum_{i=1}^{N} \lambda_i z_i(t)$ and $e_i(t) = z_i(t) - z^*(t)$. Then, we can obtain

$$\dot{e}_i(t) = -Ae_i(t) + Bg(z_i(t)) - B\sum_{k=1}^{N} \lambda_k g(z_k(t))$$

$$
\begin{aligned}
&+w_i(t) - \sum_{k=1}^{N} \lambda_k w_k(t) + c\sum_{j=1}^{N} G_{ij}\Gamma e_j(t), \\
\bar{y}_i(t) &= He_i(t),\ i = 1, 2, \cdots, N,
\end{aligned} \tag{5.32}
$$

where $\bar{y}_i(t) = y_i(t) - Hz^*(t)$.

Definition 5.10 The outputs of CNNs (5.31) are $\mathcal{H}_\infty$ synchronized with disturbance attenuation level $\gamma \in \mathbb{R} > 0$ if

$$
\sum_{i=1}^{N} \int_0^{t_p} \bar{y}_i^T(t)\bar{y}_i(t)dt \leqslant S(0) + \gamma^2 \sum_{i=1}^{N} \int_0^{t_p} w_i^T(t)w_i(t)dt
$$

for any $t_p \in \mathbb{R} > 0$ and some nonnegative function $S(\cdot)$.

Denote

$$
\begin{aligned}
\lambda &= \operatorname{diag}(\lambda_1, \lambda_2, \cdots, \lambda_N), \\
e(t) &= (e_1^T(t), e_2^T(t), \cdots, e_N^T(t))^T.
\end{aligned}
$$

Theorem 5.11 *The outputs of CNNs* (5.31) *are $\mathcal{H}_\infty$ synchronized if there exist a positive real number γ and a matrix $P \in \mathbb{R}^{nN\times nN} > 0$ such that*

$$
\begin{pmatrix} \Xi_1 & \lambda \otimes P \\ \lambda \otimes P & -\gamma^2(I_N \otimes I_n) \end{pmatrix} \leqslant 0, \tag{5.33}
$$

where $\Xi_1 = \lambda \otimes (-PA - AP + PBB^TP + \Delta) + (c\lambda G) \otimes (P\Gamma) + (cG^T\lambda) \otimes (\Gamma P) + I_N \otimes (H^TH)$.

Proof Choose the following Lyapunov functional for system (5.32):

$$
V_1(e(t)) = \sum_{i=1}^{N} \lambda_i e_i^T(t)Pe_i(t).
$$

The derivative of $V_1(e(t))$ is given by

$$
\begin{aligned}
\dot{V}_1(e(t)) \leqslant &\sum_{i=1}^{N} \lambda_i e_i^T(t)(-PA - AP + PBB^TP \\
&+\Delta)e_i(t) + 2\sum_{i=1}^{N} \lambda_i e_i^T(t)Pw_i(t)
\end{aligned}
$$

$$+2c\sum_{i=1}^{N}\sum_{j=1}^{N}\lambda_i G_{ij} e_i^T(t) P\Gamma e_j(t). \tag{5.34}$$

By (5.33) and (5.34), one has

$$\begin{aligned}
&\dot{V}_1(e(t)) + \sum_{i=1}^{N}\bar{y}_i^T(t)\bar{y}_i(t) - \gamma^2\sum_{i=1}^{N} w_i^T(t)w_i(t)\\
&\leqslant \xi^T(t)\begin{pmatrix} \Xi_1 & \lambda\otimes P \\ \lambda\otimes P & -\gamma^2(I_N\otimes I_n)\end{pmatrix}\xi(t)\\
&\leqslant 0,
\end{aligned} \tag{5.35}$$

where $\xi(t) = (e^T(t), w^T(t))^T$.

From (5.35), we can get

$$\sum_{i=1}^{N}\int_0^{t_p}\bar{y}_i^T(t)\bar{y}_i(t)dt \leqslant V_1(e(0)) + \gamma^2\sum_{i=1}^{N}\int_0^{t_p} w_i^T(t)w_i(t)dt$$

for any $t_p \in \mathbb{R} > 0$. Therefore, the outputs of CNNs (5.31) are $\mathcal{H}_\infty$ synchronized. □

5.3.2.2 Undirected Topology

When the topological structure of CNNs (5.31) is undirected, G becomes a symmetric matrix. Then, one has

$$(1, 1, \cdots, 1)G = 0.$$

Define $z^*(t) = \frac{1}{N}\sum_{i=1}^{N} z_i(t)$ and $e_i(t) = z_i(t) - z^*(t)$. Then, we can obtain

$$\begin{aligned}
\dot{e}_i(t) &= -Ae_i(t) + Bg(z_i(t)) - B\sum_{k=1}^{N}\frac{g(z_k(t))}{N}\\
&\quad + w_i(t) - \sum_{k=1}^{N}\frac{w_k(t)}{N} + c\sum_{j=1}^{N} G_{ij}\Gamma e_j(t),\\
\bar{y}_i(t) &= He_i(t),\ i = 1, 2, \cdots, N,
\end{aligned} \tag{5.36}$$

where $\bar{y}_i(t) = y_i(t) - Hz^*(t)$.

Theorem 5.12 *The outputs of CNNs* (5.31) *are $\mathcal{H}_\infty$ synchronized if there exist a positive real number γ and a matrix $P \in \mathbb{R}^{nN\times nN} > 0$ such that*

$$\begin{pmatrix} \Xi_2 & I_N \otimes P \\ I_N \otimes P & -\gamma^2(I_N \otimes I_n) \end{pmatrix} \leqslant 0, \tag{5.37}$$

where $\Xi_2 = I_N \otimes (-PA - AP + PBB^T P + \Delta) + (cG) \otimes (P\Gamma + \Gamma P) + I_N \otimes (H^T H)$.

Proof Choose the following Lyapunov functional for system (5.36):

$$V_2(e(t)) = \sum_{i=1}^{N} e_i^T(t) P e_i(t),$$

where $e(t) = (e_1^T(t), e_2^T(t), \cdots, e_N^T(t))^T$.

Then, one obtains

$$\begin{aligned} \dot{V}_2(e(t)) \leqslant & \sum_{i=1}^{N} e_i^T(t)(-PA - AP + PBB^T P + \Delta)e_i(t) \\ & +2\sum_{i=1}^{N} e_i^T(t) P w_i(t) \\ & +2c \sum_{i=1}^{N}\sum_{j=1}^{N} G_{ij} e_i^T(t) P\Gamma e_j(t). \end{aligned} \tag{5.38}$$

By (5.37) and (5.38), one has

$$\begin{aligned} & \dot{V}_2(e(t)) + \sum_{i=1}^{N} \bar{y}_i^T(t)\bar{y}_i(t) - \gamma^2 \sum_{i=1}^{N} w_i^T(t) w_i(t) \\ \leqslant & \ \xi^T(t) \begin{pmatrix} \Xi_2 & I_N \otimes P \\ I_N \otimes P & -\gamma^2(I_N \otimes I_n) \end{pmatrix} \xi(t) \\ \leqslant & \ 0, \end{aligned} \tag{5.39}$$

where $\xi(t) = (e^T(t), w^T(t))^T$.

From (5.39), we can get

$$\sum_{i=1}^{N} \int_0^{t_p} \bar{y}_i^T(t)\bar{y}_i(t)dt \leqslant V_2(e(0)) + \gamma^2 \sum_{i=1}^{N} \int_0^{t_p} w_i^T(t) w_i(t) dt$$

for any $t_p \in \mathbb{R} > 0$. Therefore, the outputs of CNNs (5.31) are $\mathcal{H}_\infty$ synchronized. □

5.3.3 $\mathcal{H}_\infty$ Output Synchronization of CNNs with Adaptive Coupling

5.3.3.1 Directed Topology

In view of Lemma 1.10, we can find a positive vector $(\varrho_1, \varrho_2, \cdots, \varrho_N)^T \in \mathbb{R}^N$ satisfying

$$(\varrho_1, \varrho_2, \cdots, \varrho_N)G(0) = 0.$$

Take

$$\lambda_i = \frac{\varrho_i}{\varrho_1 + \varrho_2 + \cdots + \varrho_N}.$$

Then, one has

$$\sum_{j=1}^{N} U_{ij} = \sum_{j=1}^{N} U_{ji} = 0,$$

where $\lambda = \mathrm{diag}(\lambda_1, \lambda_2, \cdots, \lambda_N)$, $U = (U_{ij})_{N\times N} = \lambda G(0) + G^T(0)\lambda$.

In order to guarantee that the outputs of CNNs (5.31) are $\mathcal{H}_\infty$ synchronized, the following adaptive strategy for updating the coupling weights is designed:

$$\dot{G}_{ij}(t) = \begin{cases} \varpi_{ij}\lambda_j(z_i(t) - z_j(t))^T \Gamma(z_i(t) - z_j(t)), \\ \qquad\qquad \text{if } (i, j) \in \mathcal{E} \text{ and } (j, i) \in \mathcal{E}, \\ -\sum_{\substack{k=1 \\ k\neq i}}^{N} \dot{G}_{ik}(t), \qquad\qquad \text{if } i = j, \\ 0, \qquad\qquad \text{otherwise}, \end{cases} \tag{5.40}$$

where $\varpi_{ij} = \varpi_{ji}$ are positive real numbers, $\mathcal{E} \subset \{1, 2, \cdots, N\} \times \{1, 2, \cdots, N\}$ denotes the set of directed edges in CNNs (5.31).

From (5.40), one has

$$\sum_{j=1}^{N} \dot{G}_{ij}(t) = 0, \ i = 1, 2, \cdots, N,$$

$$\sum_{j=1}^{N} \lambda_j \dot{G}_{ji}(t) = \sum_{j=1}^{N} \lambda_i \dot{G}_{ij}(t) = 0, \ i = 1, 2, \cdots, N$$

for any $t \in [0, +\infty)$.

Since $(\lambda_1, \lambda_2, \cdots, \lambda_N)G(0) = 0$, $G_{ii}(0) = -\sum_{\substack{j=1\\ j\neq i}}^N G_{ij}(0)$ and $\sum_{j=1}^N U_{ij} = 0$, we get

$$\sum_{j=1}^N \lambda_j G_{ji}(t) = 0,\ i = 1, 2, \cdots, N,$$

$$G_{ii}(t) = -\sum_{\substack{j=1\\ j\neq i}}^N G_{ij}(t),\ i = 1, 2, \cdots, N,$$

$$\sum_{j=1}^N \Big(\lambda_i G_{ij}(t) + \lambda_j G_{ji}(t)\Big) = 0$$

for any $t \in [0, +\infty)$.

Define $z^*(t) = \sum_{i=1}^N \lambda_i z_i(t)$ and $e_i(t) = z_i(t) - z^*(t)$. Then, we can get

$$\begin{aligned}\dot{e}_i(t) &= -Ae_i(t) + Bg(z_i(t)) - B\sum_{k=1}^N \lambda_k g(z_k(t)) \\ &\quad + w_i(t) - \sum_{k=1}^N \lambda_k w_k(t) + c\sum_{j=1}^N G_{ij}(t)\Gamma e_j(t), \\ \bar{y}_i(t) &= He_i(t),\end{aligned} \tag{5.41}$$

where $\bar{y}_i(t) = y_i(t) - Hz^*(t), i = 1, 2, \cdots, N$.

Denote

$$F_{ij} = \begin{cases} \lambda_i G_{ij}(0), & \text{if } (i, j) \in \mathcal{E} \text{ and } (j, i) \notin \mathcal{E}, \\ \lambda_j G_{ji}(0), & \text{if } (j, i) \in \mathcal{E} \text{ and } (i, j) \notin \mathcal{E}, \\ -\sum_{\substack{k=1\\ k\neq i}}^N F_{ik}, & \text{if } i = j, \\ 0, & \text{otherwise.} \end{cases}$$

Theorem 5.13 *The outputs of CNNs* (5.31) *are $\mathcal{H}_\infty$ synchronized under the adaptive strategy* (5.40) *if there exist a positive real number γ and a matrix $M = (M_{ij})_{N\times N} \in \mathbb{R}^{N\times N}$ such that*

$$\begin{pmatrix} \Xi_3 & \lambda \otimes I_n \\ \lambda \otimes I_n & -\gamma^2(I_N \otimes I_n) \end{pmatrix} \leqslant 0, \tag{5.42}$$

where $\Xi_3 = \lambda \otimes (-2A + BB^T + \Delta) + c(F + M) \otimes \Gamma + I_N \otimes (H^T H)$, $F = (F_{ij})_{N\times N} \in \mathbb{R}^{N\times N}$, $M_{ij} = M_{ji} (i \neq j)$ *are nonnegative real numbers,* $M_{ii} = -\sum_{\substack{j=1 \\ j\neq i}}^{N} M_{ij}$, *and* $M_{ij} = 0 (i \neq j)$ *if and only if* $G_{ij}(0) \times G_{ji}(0) = 0$.

Proof Choose the following Lyapunov functional for system (5.41):

$$
\begin{aligned}
V_3(e(t)) = &\sum_{i=1}^{N} \lambda_i e_i^T(t) e_i(t) \\
&+ c \sum_{i=1}^{N} \sum_{\substack{(i,j)\in\mathcal{E}, \\ (j,i)\in\mathcal{E}}} \frac{(\lambda_i G_{ij}(t) + \lambda_j G_{ji}(t) - M_{ij})^2}{8\varpi_{ij}\lambda_i\lambda_j},
\end{aligned}
$$

where $e(t) = (e_1^T(t), e_2^T(t), \cdots, e_N^T(t))^T$.

Then, one gets

$$
\begin{aligned}
\dot{V}_3(e(t)) \leqslant &\sum_{i=1}^{N} \lambda_i e_i^T(t)(-2A + BB^T + \Delta) e_i(t) \\
&+ 2\sum_{i=1}^{N} \lambda_i e_i^T(t) w_i(t) \\
&+ c\sum_{i=1}^{N}\sum_{j=1}^{N} (F_{ij} + M_{ij}) e_i^T(t) \Gamma e_j(t).
\end{aligned} \tag{5.43}
$$

By (5.42) and (5.43), one has

$$
\begin{aligned}
&\dot{V}_3(e(t)) + \sum_{i=1}^{N} \bar{y}_i^T(t)\bar{y}_i(t) - \gamma^2 \sum_{i=1}^{N} w_i^T(t) w_i(t) \\
&\leqslant \xi^T(t) \begin{pmatrix} \Xi_3 & \lambda \otimes I_n \\ \lambda \otimes I_n & -\gamma^2 (I_N \otimes I_n) \end{pmatrix} \xi(t) \\
&\leqslant 0,
\end{aligned} \tag{5.44}
$$

where $\xi(t) = (e^T(t), w^T(t))^T$.

From (5.44), we can get

$$
\sum_{i=1}^{N} \int_0^{t_p} \bar{y}_i^T(t)\bar{y}_i(t) dt \leqslant V_3(e(0)) + \gamma^2 \sum_{i=1}^{N} \int_0^{t_p} w_i^T(t) w_i(t) dt
$$

for any $t_p \in \mathbb{R} > 0$. Therefore, the outputs of CNNs (5.31) are $\mathcal{H}_\infty$ synchronized under the adaptive strategy (5.40). □

5.3.3.2 Undirected Topology

When the CNNs (5.31) is undirected, $G(0)$ becomes a symmetric matrix. Then, one has

$$(1, 1, \cdots, 1)G(0) = 0.$$

In order to guarantee that the outputs of CNNs (5.31) are $\mathcal{H}_\infty$ synchronized, the following adaptive strategy for updating the coupling weights is designed:

$$\dot{G}_{ij}(t) = \begin{cases} \varpi_{ij}(z_i(t) - z_j(t))^T \Gamma(z_i(t) - z_j(t)), & \\ \qquad \text{if } (i, j) \in \mathcal{E}, & \\ -\sum_{\substack{k=1 \\ k \neq i}}^N \dot{G}_{ik}(t), & \text{if } i = j, \\ 0, & \text{otherwise}, \end{cases} \tag{5.45}$$

where $\varpi_{ij} = \varpi_{ji}$ are positive real numbers, $\mathcal{E}$ denotes the set of undirected edges in CNNs (5.31).

Define $z^*(t) = \frac{1}{N}\sum_{i=1}^N z_i(t)$ and $e_i(t) = z_i(t) - z^*(t)$. Then, we can obtain

$$\begin{aligned} \dot{e}_i(t) &= -Ae_i(t) + Bg(z_i(t)) - B\sum_{k=1}^N \frac{g(z_k(t))}{N} \\ &\quad + w_i(t) - \sum_{k=1}^N \frac{w_k(t)}{N} + c\sum_{j=1}^N G_{ij}(t)\Gamma e_j(t), \\ \bar{y}_i(t) &= He_i(t), \; i = 1, 2, \cdots, N, \end{aligned} \tag{5.46}$$

where $\bar{y}_i(t) = y_i(t) - Hz^*(t)$.

Theorem 5.14 *The outputs of CNNs* (5.31) *are $\mathcal{H}_\infty$ synchronized under the adaptive strategy* (5.45).

Proof Choose the following Lyapunov functional for system (5.46):

$$\begin{aligned} V_4(e(t)) &= \sum_{i=1}^N e_i^T(t)e_i(t) \\ &\quad + c\sum_{i=1}^N \sum_{(i,j)\in\mathcal{E}} \frac{(G_{ij}(t) - M_{ij})^2}{2\varpi_{ij}}, \end{aligned}$$

where $e(t) = (e_1^T(t), e_2^T(t), \cdots, e_N^T(t))^T$, $M_{ij} = M_{ji}(i \neq j)$ are nonnegative real numbers, $M_{ii} = -\sum_{\substack{j=1 \\ j\neq i}}^{N} M_{ij}$, and $M_{ij} = 0(i \neq j)$ if and only if $G_{ij}(0) = 0$.

Then, one has

$$\dot{V}_4(e(t)) \leqslant e^T(t)\Big[I_N \otimes (-2A + BB^T + \Delta) + 2cM \otimes \Gamma\Big]e(t) + 2e^T(t)(I_N \otimes I_n)w(t),$$

where $M = (M_{ij})_{N\times N} \in \mathbb{R}^{N\times N}$.

Thus,

$$\dot{V}_4(e(t)) + \sum_{i=1}^{N} \bar{y}_i^T(t)\bar{y}_i(t) - \gamma^2 \sum_{i=1}^{N} w_i^T(t)w_i(t)$$
$$\leqslant e^T(t)\Big[I_N \otimes (-2A + BB^T + \Delta + H^T H + \frac{1}{\gamma^2}I_n) + 2cM \otimes \Gamma\Big]e(t),$$

where $0 < \gamma \in \mathbb{R}$.

Similar to the proof of Theorem 5.7, we can obtain

$$\dot{V}_4(e(t)) + \sum_{i=1}^{N} \bar{y}_i^T(t)\bar{y}_i(t) - \gamma^2 \sum_{i=1}^{N} w_i^T(t)w_i(t) \leqslant 0. \tag{5.47}$$

From (5.47), we can get

$$\sum_{i=1}^{N} \int_0^{t_p} \bar{y}_i^T(t)\bar{y}_i(t)dt \leqslant V_4(e(0)) + \gamma^2 \sum_{i=1}^{N} \int_0^{t_p} w_i^T(t)w_i(t)dt$$

for any $\mathbb{R} \ni t_p > 0$. Therefore, the outputs of CNNs (5.31) are $\mathcal{H}_\infty$ synchronized under the adaptive strategy (5.45). □

Remark 5.15 In subsection 5.3.2, we respectively analyze $\mathcal{H}_\infty$ output synchronization of directed and undirected CNNs (5.31). However, in many circumstances, the outputs of CNNs with given coupling weights are not $\mathcal{H}_\infty$ synchronized. Therefore, two adaptive strategies for updating the coupling weights are also developed in subsection 5.3.3 (see (5.40) and (5.45)). By employing these designed adaptive strategies, we establish two criteria to guarantee that the outputs of CNNs (5.31) are $\mathcal{H}_\infty$ synchronized in the scenarios with directed and undirected topologies (see Theorems 5.13 and 5.14).

5.4 Numerical Examples

Example 5.16 Consider the following directed CNNs:

$$\dot{z}_i(t) = -Az_i(t) + Bg(z_i(t)) + J + c\sum_{j=1}^{5} G_{ij}(t)\Gamma z_j(t),$$
$$y_i(t) = Hz_i(t), i = 1, 2, \cdots, 5,$$

where $c = 0.8$, $g_k(\varphi) = \frac{1}{4}(|\varphi+1| - |\varphi - 1|)$, $k = 1, 2, 3$, $A = \text{diag}(0.8, 0.8, 0.9)$, $\Gamma = \text{diag}(0.7, 0.5, 0.8)$, $J = \text{diag}(0, 0, 0)$, and the matrices H, B, $G(0)$ are chosen as, respectively,

$$H = \begin{pmatrix} 1 & 0 & 0 \\ 0 & 1 & 0 \end{pmatrix},$$

$$B = \begin{pmatrix} 0.2 & 0.2 & 0.2 \\ 0.1 & 0.1 & 0.2 \\ 0.3 & 0.1 & 0.2 \end{pmatrix},$$

$$G(0) = \begin{pmatrix} -0.25 & 0.25 & 0 & 0 & 0 \\ 0 & -0.1 & 0 & 0.1 & 0 \\ 0 & 0 & -0.2 & 0 & 0.2 \\ 0 & 0.05 & 0.25 & -0.3 & 0 \\ 0.25 & 0 & 0.05 & 0 & -0.3 \end{pmatrix}.$$

Obviously, function $g_k(\cdot)$ satisfies Lipschitz condition with $\beta_k = 0.5$. Using the MATLAB function NULL, we can find vector $(1, 3, 1.5, 1, 1)^T \in \mathbb{R}^5$ satisfying $(1, 3, 1.5, 1, 1)G(0) = 0$.

In what follows, by simple calculations, we can obtain

$$\lambda = \text{diag}\left(\frac{2}{15}, \frac{2}{5}, \frac{1}{5}, \frac{2}{15}, \frac{2}{15}\right),$$

$$F = \begin{pmatrix} -\frac{1}{15} & \frac{1}{30} & 0 & 0 & \frac{1}{30} \\ \frac{1}{30} & -\frac{1}{30} & 0 & 0 & 0 \\ 0 & 0 & -\frac{1}{30} & \frac{1}{30} & 0 \\ 0 & 0 & \frac{1}{30} & -\frac{1}{30} & 0 \\ \frac{1}{30} & 0 & 0 & 0 & -\frac{1}{30} \end{pmatrix}.$$

We can find the following:

$$k = 0.1868 > 0,$$

$$M = \begin{pmatrix} 0 & 0 & 0 & 0 & 0 \\ 0 & -0.6539 & 0 & 0.6539 & 0 \\ 0 & 0 & -0.7390 & 0 & 0.7390 \\ 0 & 0.6539 & 0 & -0.6539 & 0 \\ 0 & 0 & 0.7390 & 0 & -0.7390 \end{pmatrix}$$

satisfying the condition of Theorem 5.6. According to Theorem 5.6, the CNNs (5.1) with above given parameters achieves output synchronization under the adaptive strategy (5.21). Take $\varpi_{24} = \varpi_{42} = 1$, $\varpi_{35} = \varpi_{53} = 1$, and the simulation results are shown in Figs. 5.1 and 5.2.

Remark 5.17 By utilizing the Matlab, we can easily get the full rank decomposition for any matrix $H \in \mathbb{R}^{q\times n}$:

$$H = \tilde{H}_1 \tilde{H}_2,$$

where $r = \text{Rank}(H)$, $\tilde{H}_1 \in \mathbb{R}^{q\times r}$ and $\tilde{H}_2 \in \mathbb{R}^{r\times n}$. Obviously, $\tilde{H}_1 \in \mathbb{R}^{q\times r}$ can be transformed into $\begin{pmatrix} I_r \\ 0 \end{pmatrix}_{q\times r}$ by a series of simple row transformations, and $\tilde{H}_2 \in \mathbb{R}^{r\times n}$ can be transformed into $(I_r \quad 0)_{r\times n}$ by a series of simple column transformations. Namely, we can find non-singular matrices $\bar{H}_1 \in \mathbb{R}^{q\times q}$ and $\bar{H}_2 \in$

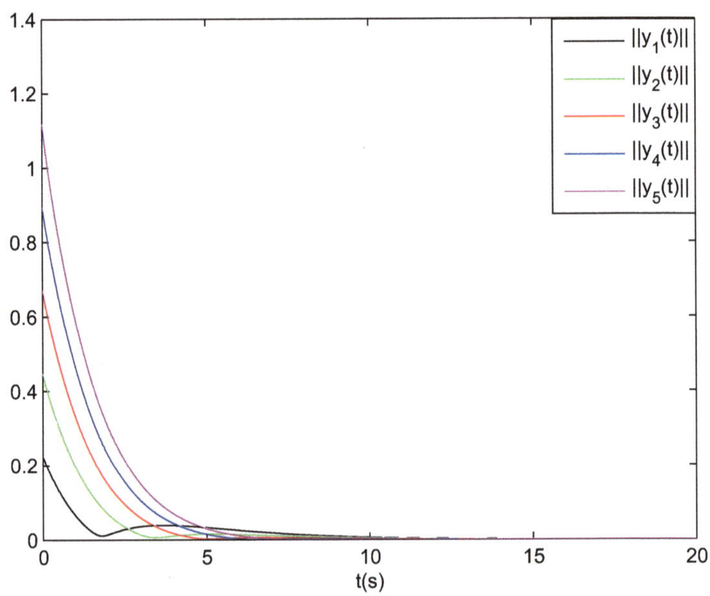

Fig. 5.1 The norms of the output vectors $y_i(t), i = 1, 2, \cdots, 5$

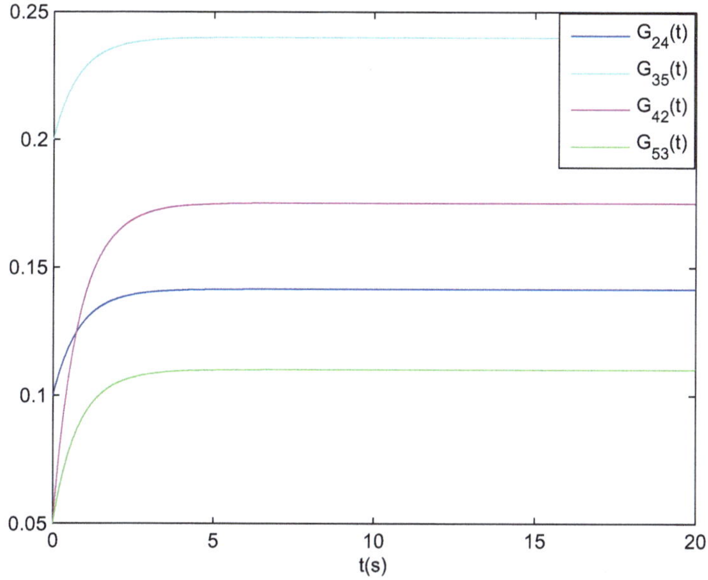

Fig. 5.2 The adaptive coupling weights

$\mathbb{R}^{n\times n}$ such that $\tilde{H}_1 = \bar{H}_1 \begin{pmatrix} I_r \\ 0 \end{pmatrix}_{q\times r}$ and $\tilde{H}_2 = (I_r \ \ 0)_{r\times n} \bar{H}_2$. Letting $H_1 = \bar{H}_1$ and $H_2 = \bar{H}_2$, one has $H = H_1 \begin{pmatrix} I_r \\ 0 \end{pmatrix}_{q\times r} (I_r \ \ 0)_{r\times n} H_2 = H_1 \begin{pmatrix} I_r & 0 \\ 0 & 0 \end{pmatrix}_{q\times n} H_2$ and $H_3 = (I_r \ \ 0)_{r\times n} H_2 = \tilde{H}_2$. From the proof of Theorems 5.3–5.7, it can be found that we only need to compute H_3.

Example 5.18 Consider the following directed CNNs with external disturbances:

$$\begin{aligned} \dot{z}_i(t) &= -Az_i(t) + Bg(z_i(t)) + J + w_i(t) \\ &\quad + c\sum_{j=1}^{5} G_{ij}(t)\Gamma z_j(t), \\ y_i(t) &= Hz_i(t), \ \ i = 1, 2, \cdots, 5, \end{aligned}$$

where c=0.7, $g_k(\varphi)$=$\frac{1}{4}(|\varphi+1| - |\varphi-1|)$, k=1, 2, 3, A = diag(0.6, 0.8, 0.7), Γ = diag(0.8, 0.9, 0.7), $J = (0, 0, 0)^T$, and the matrices $H, B, G(0)$ are chosen as, respectively,

$$H = \begin{pmatrix} 0.1 & 0 & 0 \\ 0 & 0.1 & 0 \end{pmatrix},$$

$$B = \begin{pmatrix} 0.2 & 0.1 & 0.1 \\ 0.3 & 0.4 & 0.2 \\ 0.5 & 0.6 & 0.1 \end{pmatrix}.$$

$$G(0) = \begin{pmatrix} -0.8 & 0.8 & 0 & 0 & 0 \\ 0 & -0.6 & 0 & 0.6 & 0 \\ 0 & 0 & -0.5 & 0 & 0.5 \\ 0 & 0.2 & 0.4 & -0.6 & 0 \\ 0.8 & 0 & 0.1 & 0 & -0.9 \end{pmatrix}.$$

Obviously, function $g_k(\cdot)$ satisfies Lipschitz condition with $\beta_k = 0.5$. Using the MATLAB function NULL, we can find vector $(1, 2, 1.8, 2, 1)^T \in \mathbb{R}^5$ satisfying $(1, 2, 1.8, 2, 1)G(0) = 0$.

In what follows, by simple calculations, we can obtain

$$\lambda = \text{diag}\left(\frac{5}{39}, \frac{10}{39}, \frac{3}{13}, \frac{10}{39}, \frac{5}{39}\right),$$

$$F = \begin{pmatrix} -\frac{8}{39} & \frac{4}{39} & 0 & 0 & \frac{4}{39} \\ \frac{4}{39} & -\frac{4}{39} & 0 & 0 & 0 \\ 0 & 0 & -\frac{4}{39} & \frac{4}{39} & 0 \\ 0 & 0 & \frac{4}{39} & -\frac{4}{39} & 0 \\ \frac{4}{39} & 0 & 0 & 0 & -\frac{4}{39} \end{pmatrix}.$$

We can find the following:

$$\gamma = 1.2190 > 0,$$

$$M = \begin{pmatrix} 0 & 0 & 0 & 0 & 0 \\ 0 & -1.1712 & 0 & 1.1712 & 0 \\ 0 & 0 & -1.2007 & 0 & 1.2007 \\ 0 & 1.1712 & 0 & -1.1712 & 0 \\ 0 & 0 & 1.2007 & 0 & -1.2007 \end{pmatrix}$$

satisfying the condition of Theorem 5.13.

According to Theorem 5.13, the CNNs (5.31) with above given parameters achieves $\mathcal{H}_\infty$ output synchronization under the adaptive strategy (5.40). The simulation results are shown in Figs. 5.3 and 5.4.

Figs. 5.1 and 5.2 show the norms of the output vectors $y_i(t), i = 1, 2, \cdots, 5$, and the adaptive coupling weights for CNNs (5.1) under the adaptive law (5.21). As shown in the figures, the output vectors $y_i(t), i = 1, 2, \cdots, 5$ are synchronized and the coupling weights $G_{24}(t), G_{35}(t), G_{42}(t), G_{53}(t)$ gradually converge to finite values. In Figs. 5.3 and 5.4, the curves of the norms of the vectors $e_i(t), \bar{y}_i(t)$,

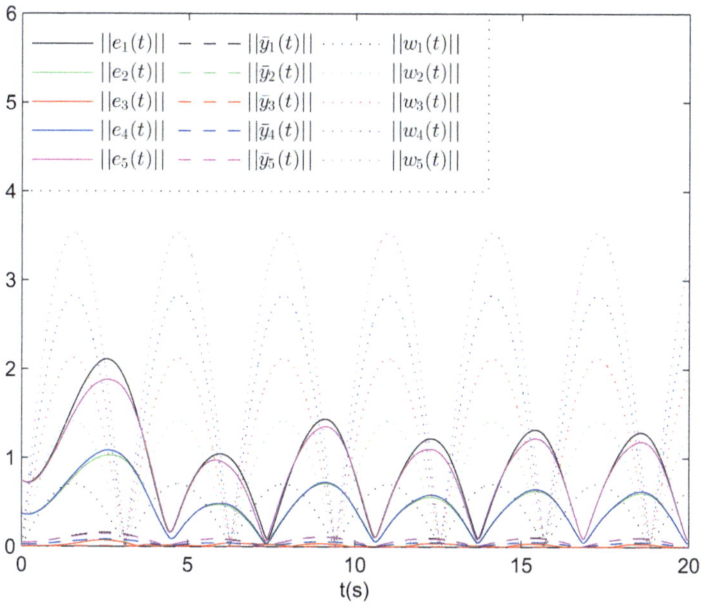

Fig. 5.3 The norms of the vectors $e_i(t)$, $\bar{y}_i(t)$, and $w_i(t)$, $i = 1, 2, \cdots, 5$

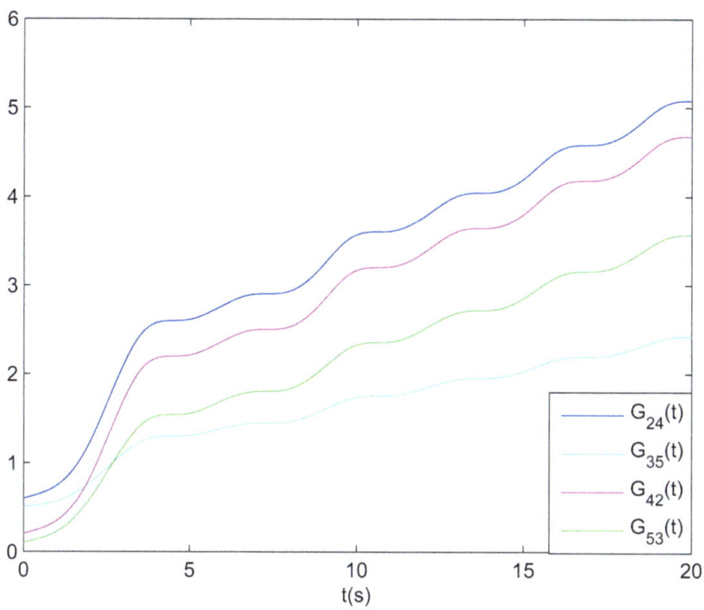

Fig. 5.4 The adaptive coupling weights

$w_i(t), i = 1, 2, \cdots, 5$, and the adaptive coupling weights of CNNs (5.31) under the adaptive law (5.40), respectively, are plotted.

Remark 5.19 Practically, the parameter ϖ_{ij} in adaptive strategies (5.21), (5.28), (5.40) and (5.45) can be selected as any positive real numbers. But, in order to reduce simulation errors, it may be better to choose the value of ϖ_{ij} as small as possible.

5.5 Conclusion

The output synchronization and $\mathcal{H}_\infty$ output synchronization problems for directed and undirected CNNs have been investigated. By employing matrix theory, Lyapunov functionals and Barbalat's Lemma, several output synchronization and $\mathcal{H}_\infty$ output synchronization criteria have been presented for these network models. Moreover, some adaptive schemes to adjust coupling weights have been developed for reaching output synchronization and $\mathcal{H}_\infty$ output synchronization. Two numerical examples have also been presented to show the effectiveness of the proposed adaptive schemes.

Chapter 6
Local and Global Exponential Output Synchronization of CDDNs

6.1 Introduction

Practically, there are two kinds of coupling forms in complex networks: state coupling and output coupling. Synchronization [7, 22, 23, 58, 59, 62, 65, 66, 150–156] in complex networks has been investigated for many years. However, there are few works on the synchronization of complex networks with output coupling.

To our knowledge, Jiang, Tang and Chen [22] first introduced a complex network model with output coupling without time delays. Some conditions for synchronization were established based on the Lyapunov stability theory. In [23], Chen proposed a complex network model with output coupling. A sufficient synchronization condition was given to ensure that the proposed network model is exponentially mean-square stable. It should be noticed that the state synchronization of complex networks was considered in these articles. However, the node state in complex networks is difficult to be observed or measured, even the node state can not be observed or measured at all. Moreover, in many circumstances only part states are needed to make the synchronization to come true. Therefore, it is interesting to study the output synchronization of complex networks with output coupling.

Motivated by the above discussions, we propose a new CDDN model. The objective of this chapter is to study the output synchronization of the proposed network model. Some sufficient conditions ensuring local and global exponential output synchronization are obtained by Lyapunov functional method.

6.2 Network Model

Throughout this chapter, we consider a CDDN consisting of N identical nodes with diffusive and output coupling. The mathematical model of the coupled network can be described as follows [22, 23, 58]:

J.-L. Wang et al., *Analysis and Control of Output Synchronization for Complex Dynamical Networks*, https://doi.org/10.1007/978-981-13-1352-3_6

$$\begin{cases} \dot{x}_i(t) = f(x_i(t)) + \dfrac{a}{k_i^{\beta_\omega}} \sum\limits_{j=1}^{N} L_{ij} \Gamma_1 y_j(t) + \dfrac{b}{k_i^{\beta_\omega}} \sum\limits_{j=1}^{N} L_{ij} \Gamma_2 y_j(t - \tau(t)), \\ y_i(t) = C x_i(t), \end{cases} \tag{6.1}$$

where $i = 1, 2, \cdots, N$. $\tau(t)$ is the time-varying delay with $0 \leqslant \tau(t) \leqslant \tau$.

The function $f(\cdot)$, describing the local dynamics of the nodes, is continuously differentiable and capable of producing various rich dynamical behaviors, $x_i(t) = (x_{i1}(t), x_{i2}(t), \cdots, x_{in}(t))^T \in \mathbb{R}^n$ is the state variable of node i, $y_i(t) \in \mathbb{R}^q$ $(1 \leqslant q \leqslant n)$ is the output vector of node i, C is known matrix with appropriate dimensions, $\Gamma_1, \Gamma_2 \in \mathbb{R}^{n\times q}$ are inner-coupling matrix, which describe the individual coupling between two nodes for non-delayed configuration and delayed one, respectively, $a, b > 0$ represent the coupling strength for non-delayed configuration and delayed one, respectively, k_i is the degree of node i and β_ω is a tunable weight parameter, the real matrix $L = (L_{ij})_{N\times N}$ is a symmetric matrix with diagonal entries $L_{ii} = -k_i$ and off-diagonal entries $L_{ij} = 1$ if node i and node j are connected by a link, and $L_{ij} = 0$ otherwise.

In this chapter, we always assume that complex dynamical network (6.1) is connected. Let

$$\begin{aligned} x(t) &= (x_1^T(t),\ x_2^T(t),\ \cdots,\ x_N^T(t))^T, \\ y(t) &= (y_1^T(t), y_2^T(t), \cdots, y_N^T(t))^T. \end{aligned}$$

The initial condition associated with the complex dynamical network (6.1) is given in form

$$\begin{aligned} \Phi(s) &= (\phi_1^T(s), \phi_2^T(s), \cdots, \phi_N^T(s))^T, \\ x(t_0 + s) &= \Phi(s),\ \ \phi_i \in C([-\tau, 0], \mathbb{R}^n), \\ y(t_0 + s) &= \mathrm{diag}(C, C, \cdots, C)\Phi(s). \end{aligned}$$

Remark 6.1 In this model, the parameter $\beta_\omega = 0$ recovers that the network is unweighted and undirected, and the condition $\beta_\omega \neq 0$ corresponds to a network with weighted configuration and bidirectional [58]. However, [22] and [23] only considered the case that the network is undirected and/or unweighted, and time delay wasn't considered in [22]. Moreover, we take into account the case that the coupling relations and the coupling strength are related to the current output and the delayed output.

Definition 6.2 The complex dynamical network (6.1) is said to achieve output synchronization if

$$\lim_{t\to+\infty} \|y_i(t) - y_j(t)\| = 0, \text{ for all } i, j = 1, 2, ..., N.$$

6.3 Main Results

For convenient analysis, we let

$$G_{ij} = \frac{L_{ij}}{k_i^{\beta_\omega}},$$

then network (6.1) can be rewritten as

$$\begin{cases} \dot{x}_i(t) = f(x_i(t)) + a\sum_{j=1}^{N} G_{ij}\Gamma_1 y_j(t) + b\sum_{j=1}^{N} G_{ij}\Gamma_2 y_j(t-\tau(t)), \\ y_i(t) = Cx_i(t), \end{cases} \tag{6.2}$$

where $i = 1, 2, \cdots, N$. $G = (G_{ij})_{N\times N}$ is a coupling matrix, accounting for the topology of complex network.

Since network (6.1) is connected in the sense of having no isolated clusters, which means that the matrix K ($K = \text{diag}(k_1, k_2, \cdots, k_N)$ is the diagonal matrix of degree.) is non-singular, and the coupling matrix L is irreducible. It is obvious that we can obtain $G = K^{-\beta_\omega}L$. Using matrix identities, we know that the spectrum of asymmetric matrix G is equal to the spectrum of the symmetric matrix $W = K^{-\beta_\omega/2}LK^{-\beta_\omega/2}$, which is semi-negative definite. Then, we can easily obtain the eigenvalues of G are real-valued and are strictly negative except an eigenvalue 0 with multiplicity 1, and G can be diagonalizable. Hence, there exists a nonsingular matrix

$$\phi = (\phi_1, \phi_2, \cdots, \phi_N) \in R^{N\times N}$$

such that

$$\phi^{-1}G\phi = \Lambda$$

with

$$\Lambda = \text{diag}(\lambda_1, \lambda_2, \cdots, \lambda_N).$$

$\lambda_i, i = 1, 2, \ldots, N$, are the eigenvalues of G and

$$0 = \lambda_1 > \lambda_2 \geqslant \lambda_3 \geqslant \cdots \geqslant \lambda_N.$$

On the other hand, there obviously exist nonsingular matrices $U \in \mathbb{R}^{q\times q}$ and $H \in \mathbb{R}^{n\times n}$ such that

$$C = U \begin{pmatrix} I_r & 0 \\ 0 & 0 \end{pmatrix} H,$$

where $r = \text{Rank}(C)$.

Set

$$\hat{x}_i(t) = Hx_i(t),$$

and

$$\hat{y}_i(t) = U^{-1} y_i(t).$$

System (6.2) can be rewritten as

$$\begin{cases} \dot{\hat{x}}_i(t) = Hf(H^{-1}\hat{x}_i(t)) + a\sum_{j=1}^{N} G_{ij} H\Gamma_1 U \hat{y}_j(t) \\ \qquad + b\sum_{j=1}^{N} G_{ij} H\Gamma_2 U \hat{y}_j(t-\tau(t)), \\ \hat{y}_i(t) = \begin{pmatrix} I_r & 0 \\ 0 & 0 \end{pmatrix} \hat{x}_i(t), \end{cases} \tag{6.3}$$

where $i = 1, 2, \cdots, N$.

Namely,

$$\dot{\hat{y}}_i(t) = \hat{f}(\hat{x}_i(t)) + a\sum_{j=1}^{N} G_{ij} \begin{pmatrix} I_r & 0 \\ 0 & 0 \end{pmatrix} H\Gamma_1 U \hat{y}_j(t) + b\sum_{j=1}^{N} G_{ij} \begin{pmatrix} I_r & 0 \\ 0 & 0 \end{pmatrix} H\Gamma_2 U \hat{y}_j(t-\tau(t)), \tag{6.4}$$

where $i = 1, 2, \cdots, N$, $\hat{f}(\hat{x}_i(t)) = \begin{pmatrix} I_r & 0 \\ 0 & 0 \end{pmatrix} Hf(H^{-1}\hat{x}_i(t))$.

In order to obtain our main results, an assumption is introduced.

Assumption 1 *Assume that*

$$\hat{f}(\hat{x}_i(t)) = \begin{pmatrix} \hat{f}_1(\hat{x}_i^1(t)) \\ 0 \end{pmatrix},$$

and $s(t) \in \mathbb{R}^r$ *is an equilibrium, a periodic trajectory or a chaotic attractor of the following system*

$$\dot{s}(t) = \hat{f}_1(s(t)),$$

where

$$\hat{x}_i^1(t) = (\hat{x}_{i1}(t),\ \hat{x}_{i2}(t), \cdots, \hat{x}_{ir}(t))^T,$$
$$\hat{x}_i^2(t) = (\hat{x}_{i(r+1)}(t), \hat{x}_{i(r+2)}(t), \cdots, \hat{x}_{in}(t))^T,$$
$$\hat{x}_i(t) = ((\hat{x}_i^1(t))^T, (\hat{x}_i^2(t))^T)^T, i = 1, 2, \cdots, N.$$

By Assumption 1, (6.3) and (6.4), we have

$$\dot{\hat{x}}_i^1(t) = \hat{f}_1(\hat{x}_i^1(t)) + a\sum_{j=1}^{N} G_{ij}M\hat{x}_j^1(t) + b\sum_{j=1}^{N} G_{ij}E\hat{x}_j^1(t-\tau(t)),$$
$$i = 1, 2, \cdots, N, \tag{6.5}$$

where

$$M = \begin{pmatrix} I_r & 0 \end{pmatrix}_{r\times n} H\Gamma_1 U \begin{pmatrix} I_r \\ 0 \end{pmatrix}_{q\times r},$$

$$E = \begin{pmatrix} I_r & 0 \end{pmatrix}_{r\times n} H\Gamma_2 U \begin{pmatrix} I_r \\ 0 \end{pmatrix}_{q\times r}.$$

It is obvious that the network (6.1) achieves global (local) exponential output synchronization if and only if system (6.5) is globally (locally) exponentially stable.

Theorem 6.3 *Let Assumption* 1 *hold and* $\dot{\tau}(t) \leqslant \sigma < 1$. *If there exist matrices* $P_i, Q_i \in \mathbb{R}^{r\times r} > 0$ *and real number* $\varepsilon_i > 0$, *such that*

$$\begin{pmatrix} W_i & b\lambda_i P_i E \\ b\lambda_i E^T P_i & -(1-\sigma)e^{-\varepsilon_i\tau} Q_i \end{pmatrix} \leqslant 0, \tag{6.6}$$

where $W_i = Q_i + \varepsilon_i P_i + P_i(D\hat{f}_1(s(t)) + a\lambda_i M) + (D\hat{f}_1(s(t)) + a\lambda_i M)^T P_i$, $D\hat{f}_1(s(t)) = \hat{f}_1'(s(t)) \in \mathbb{R}^{r\times r}$ *is the Jacobian of* $\hat{f}_1(x(t))$ *at* $s(t)$, $i = 1, 2, \cdots, N$, *then complex network* (6.1) *achieves local exponential output synchronization.*

Proof Set $e_i(t) = \hat{x}_i^1(t) - s(t),\ i = 1, 2, \cdots, N$. We can get

$$\dot{e}_i(t) = \hat{f}_1(\hat{x}_i^1(t)) - \hat{f}_1(s(t)) + a\sum_{j=1}^{N} G_{ij}Me_j(t)$$

$$+b\sum_{j=1}^{N} G_{ij}Ee_j(t-\tau(t)). \tag{6.7}$$

Its linearized system reads as

$$\begin{aligned}\dot{e}_i(t) &= D\hat{f}_1(s(t))e_i(t) + a\sum_{j=1}^{N} G_{ij}Me_j(t)\\ &+b\sum_{j=1}^{N} G_{ij}Ee_j(t-\tau(t)).\end{aligned} \tag{6.8}$$

We can rewrite networks (6.8) in a compact form as follows

$$\begin{aligned}\dot{e}(t) &= [I_N \otimes D\hat{f}_1(s(t)) + a(G\otimes M)]e(t)\\ &+b(G\otimes E)e(t-\tau(t)),\end{aligned} \tag{6.9}$$

where $e(t) = \left(e_1^T(t), e_2^T(t), \cdots, e_N^T(t)\right)^T, e(t-\tau(t)) = (e_1^T(t-\tau(t)), e_2^T(t-\tau(t)), \cdots, e_N^T(t-\tau(t)))^T$.

Take a nonsingular transform

$$\begin{aligned}(\phi^{-1}\otimes I_r)e(t) &= \nu(t)\\ &= \left(\nu_1^T(t), \nu_2^T(t), \cdots, \nu_N^T(t)\right)^T \in \mathbb{R}^{rN}.\end{aligned}$$

From (6.9), we have

$$\begin{aligned}\dot{\nu}(t) &= [I_N \otimes D\hat{f}_1(s(t)) + a(\Lambda\otimes M)]\nu(t)\\ &+b(\Lambda\otimes E)\nu(t-\tau(t))\end{aligned}$$

where $\nu(t-\tau(t)) = (\nu_1^T(t-\tau(t)), \nu_2^T(t-\tau(t)), \cdots, \nu_N^T(t-\tau(t)))^T$.

Namely,

$$\dot{\nu}_i(t) = (D\hat{f}_1(s(t)) + a\lambda_i M)\nu_i(t) + b\lambda_i E\nu_i(t-\tau(t)) \tag{6.10}$$

where $i = 1, 2, \cdots, N$.

It is obvious that system (6.5) is locally exponentially stable if the above N pieces of r-dimensional linear delayed differential equations are globally exponentially stable. Next, we shall prove that system (6.10) is globally exponentially stable.

Select the Lyapunov functional

$$V\big(\nu_i(t)\big) = {\nu_i}^T(t)P_i\nu_i(t)e^{\varepsilon_i t} + \int_{t-\tau(t)}^{t} {\nu_i}^T(\alpha)Q_i\nu_i(\alpha)e^{\varepsilon_i\alpha}d\alpha. \tag{6.11}$$

The derivative of $V\big(\nu_i(t)\big)$ along the solution of the ith ($i = 1, 2, \cdots, N$) equation in system (6.10) is

$$\begin{aligned}
\dot{V}(\nu_i(t)) &= 2\nu_i{}^T(t)P_i\dot{\nu}_i(t)e^{\varepsilon_i t} + \varepsilon_i\nu_i{}^T(t)P_i\nu_i(t)e^{\varepsilon_i t} + \nu_i{}^T(t)Q_i\nu_i(t)e^{\varepsilon_i t} \\
&\quad -(1-\dot{\tau}(t))\nu_i{}^T(t-\tau(t))Q_i\nu_i(t-\tau(t))e^{\varepsilon_i(t-\tau(t))} \\
&\leqslant 2\nu_i{}^T(t)P_i\dot{\nu}_i(t)e^{\varepsilon_i t} + \varepsilon_i\nu_i{}^T(t)P_i\nu_i(t)e^{\varepsilon_i t} + \nu_i{}^T(t)Q_i\nu_i(t)e^{\varepsilon_i t} \\
&\quad -(1-\sigma)\nu_i{}^T(t-\tau(t))Q_i\nu_i(t-\tau(t))e^{\varepsilon_i(t-\tau)} \\
&= \begin{pmatrix} \nu_i(t) \\ \nu_i\big(t-\tau(t)\big) \end{pmatrix}^T \begin{pmatrix} W_i & b\lambda_i P_i E \\ b\lambda_i E^T P_i & -(1-\sigma)e^{-\varepsilon_i\tau}Q_i \end{pmatrix} \\
&\quad \times \begin{pmatrix} \nu_i(t) \\ \nu_i\big(t-\tau(t)\big) \end{pmatrix} e^{\varepsilon_i t}.
\end{aligned}$$

According to (6.6), we can obtain $\dot{V}(\nu_i(t)) \leqslant 0$. Thus

$$V(\nu_i(t)) \leqslant V(\nu_i(t_0)).$$

From the definition of $V(\nu_i(t))$, we have

$$V(\nu_i(t)) \geqslant e^{\varepsilon_i t}\lambda_m(P_i)\|\nu_i(t)\|^2.$$

On the other hand, there obviously exists a constant $\varrho > 1$ such that

$$V(\nu_i(t_0)) \leqslant \varrho\|\psi_i\|_\tau^2 e^{\varepsilon_i t_0}$$

where ψ_i is the initial condition of $\nu_i(t)$, $\psi_i = \psi_i(\theta) \in C([-\tau, 0], \mathbb{R}^r)$, $i = 1, 2, \cdots, N$.

Then, we can get

$$\|\nu_i(t)\| \leqslant \left(\frac{\varrho}{\lambda_m(P_i)}\right)^{\frac{1}{2}} \|\psi_i\|_\tau e^{\frac{-\varepsilon_i(t-t_0)}{2}}.$$

Therefore, system (6.10) is globally exponentially stable and we thus obtain the local exponential output synchronization of complex network (6.1). The proof is completed.

In the following, we shall discuss the global exponential output synchronization of complex network (6.1).

We can rewrite system (6.7) in a compact form as follows

$$\dot{e}(t) = \hat{F}(\hat{x}^1(t)) - \hat{F}(s(t)) + a(G \otimes M)e(t) \\ +b(G \otimes E)e(t - \tau(t)), \tag{6.12}$$

where $\hat{F}(\hat{x}^1(t)) = (\hat{f}_1^T(\hat{x}_1^1(t)), \hat{f}_1^T(\hat{x}_2^1(t)), \cdots, \hat{f}_1^T(\hat{x}_N^1(t)))^T, \hat{F}(s(t)) = (\hat{f}_1^T(s(t)), \hat{f}_1^T(s(t)), \cdots, \hat{f}_1^T(s(t)))^T$.

It is obvious that system (6.5) is globally exponentially stable if the delayed differential equation (6.12) is globally exponentially stable.

Next, we make the following assumption [43, 44].

Assumption 2 *There exist a positive definite diagonal matrix $Z = \mathrm{diag}(z_1, z_2, \cdots, z_r)$ and a diagonal matrix $\Delta = \mathrm{diag}(\delta_1, \delta_2, \cdots, \delta_r)$ such that $\hat{f}_1$ satisfies the following inequality:*

$$(x - y)^T Z[\hat{f}_1(x) - \hat{f}_1(y) - \Delta(x - y)] \leqslant -\eta(x - y)^T(x - y)$$

for some $\eta > 0$ and all $x, y \in \mathbb{R}^r$.

For the convenience, we denote

$$\hat{Z} = \mathrm{diag}(Z, Z, \cdots, Z), \\ \hat{\Delta} = \mathrm{diag}(\Delta, \Delta, \cdots, \Delta).$$

Theorem 6.4 *Let Assumptions* 1 *and* 2 *hold, $\dot{\tau}(t) \leqslant \sigma < 1$. If there exist matrix $Q \in \mathbb{R}^{rN \times rN} > 0$ and a positive constant ε, such that*

$$\begin{pmatrix} W & b\hat{Z}(G \otimes E) \\ b(G \otimes E)^T \hat{Z} & -(1 - \sigma)e^{-\varepsilon\tau} Q \end{pmatrix} \leqslant 0 \tag{6.13}$$

where $W = -2\eta I_{rN} + 2\hat{Z}\hat{\Delta} + Q + \varepsilon\hat{Z} + a\hat{Z}(G \otimes M) + a(G \otimes M)^T \hat{Z}$, then complex dynamical network (6.1) *achieves global exponential output synchronization.*

Proof Construct Lyapunov functional for system (6.12) as follows

$$V(e(t)) = e^T(t)\hat{Z}e(t)e^{\varepsilon t} + \int_{t-\tau(t)}^{t} e^T(\alpha)Qe(\alpha)e^{\varepsilon\alpha}d\alpha. \tag{6.14}$$

The derivative of $V(e(t))$ along the solution of the system (6.12) is

$$\begin{aligned}\dot{V}(e(t)) &= 2e^T(t)\hat{Z}\dot{e}(t)e^{\varepsilon t} + \varepsilon e^T(t)\hat{Z}e(t)e^{\varepsilon t} + e^T(t)Qe(t)e^{\varepsilon t} \\ &\quad -(1 - \dot{\tau}(t))e^T(t - \tau(t))Qe(t - \tau(t))e^{\varepsilon(t-\tau(t))} \\ &\leqslant 2e^T(t)\hat{Z}\dot{e}(t)e^{\varepsilon t} + \varepsilon e^T(t)\hat{Z}e(t)e^{\varepsilon t} + e^T(t)Qe(t)e^{\varepsilon t}\end{aligned}$$

$$
\begin{aligned}
&-(1-\sigma)e^T(t-\tau(t))Qe(t-\tau(t))e^{\varepsilon(t-\tau)}\\
&= 2e^T(t)\hat{Z}(\hat{F}(\hat{x}^1(t)) - \hat{F}(s(t)))e^{\varepsilon t} + 2ae^T(t)\hat{Z}(G\otimes M)e(t)e^{\varepsilon t}\\
&+2be^T(t)\hat{Z}(G\otimes E)e(t-\tau(t))e^{\varepsilon t} + e^T(t)Qe(t)e^{\varepsilon t} + \varepsilon e^T(t)\hat{Z}e(t)e^{\varepsilon t}\\
&-(1-\sigma)e^T(t-\tau(t))Qe(t-\tau(t))e^{\varepsilon(t-\tau)}.
\end{aligned}
$$

According to Assumption 2, we have

$$
\begin{aligned}
&e^T(t)\hat{Z}(\hat{F}(\hat{x}^1(t)) - \hat{F}(s(t)))\\
&= \sum_{i=1}^N e_i^T(t)Z(\hat{f}_1(\hat{x}_i^1(t)) - \hat{f}_1(s(t)))\\
&= \sum_{i=1}^N (\hat{x}_i^1(t) - s(t))^T Z(\hat{f}_1(\hat{x}_i^1(t)) - \hat{f}_1(s(t)))\\
&\leqslant \sum_{i=1}^N [-\eta(\hat{x}_i^1(t) - s(t))^T(\hat{x}_i^1(t) - s(t)) + (\hat{x}_i^1(t) - s(t))^T Z\Delta(\hat{x}_i^1(t) - s(t))]\\
&= e^T(t)(-\eta I_{rN} + \hat{Z}\hat{\Delta})e(t).
\end{aligned}
\tag{6.15}
$$

It follows from (6.13) and (6.15) that

$$
\begin{aligned}
\dot{V}(e(t)) &\leqslant e^T(t)(-2\eta I_{rN} + 2\hat{Z}\hat{\Delta})e(t)e^{\varepsilon t} + 2ae^T(t)\hat{Z}(G\otimes M)e(t)e^{\varepsilon t}\\
&+2be^T(t)\hat{Z}(G\otimes E)e(t-\tau(t))e^{\varepsilon t} + e^T(t)Qe(t)e^{\varepsilon t} + \varepsilon e^T(t)\hat{Z}e(t)e^{\varepsilon t}\\
&-(1-\sigma)e^T(t-\tau(t))Qe(t-\tau(t))e^{\varepsilon(t-\tau)}\\
&= \begin{pmatrix} e(t) \\ e(t-\tau(t)) \end{pmatrix}^T \begin{pmatrix} W & b\hat{Z}(G\otimes E) \\ b(G\otimes E)^T\hat{Z} & -(1-\sigma)e^{-\varepsilon\tau}Q \end{pmatrix} \begin{pmatrix} e(t) \\ e(t-\tau(t)) \end{pmatrix} e^{\varepsilon t}\\
&\leqslant 0.
\end{aligned}
$$

Then, we can get $V(e(t)) \leqslant V(e(t_0))$. Similarly, we can conclude that the system (6.12) is globally exponentially stable. Therefore, complex dynamical network (6.1) achieves global exponential output synchronization. The proof is completed.

6.4 Examples

In this section, two illustrative examples are provided to verify the effectiveness of the proposed theoretical results.

Example 6.5 Consider a complex dynamical network, in which each node is a 3-dimensional nonlinear system described by

$$\begin{pmatrix} \dot{x}_1 \\ \dot{x}_2 \\ \dot{x}_3 \end{pmatrix} = \begin{pmatrix} -4x_1 \\ 3x_1^2 - 5x_2 \\ 4x_3 + 4x_1x_2 \end{pmatrix}.$$

Take

$$\Gamma_1 = \begin{pmatrix} 0.2 & 0.5 \\ 0.1 & 0.2 \\ 0.3 & 0.1 \end{pmatrix},$$

$$\Gamma_2 = \begin{pmatrix} 0.3 & 0.5 \\ 0.1 & 0.4 \\ 0.2 & 0.5 \end{pmatrix},$$

$$C = \begin{pmatrix} 1 & 0 & 0 \\ 0 & 1 & 0 \end{pmatrix},$$

$a = 0.1, b = 0.2$ and $\beta_\omega = 1$. The matrix L is chosen as

$$\begin{pmatrix} -3 & 1 & 1 & 0 & 0 & 0 & 0 & 0 & 0 & 1 \\ 1 & -4 & 1 & 0 & 1 & 0 & 1 & 0 & 0 & 0 \\ 1 & 1 & -5 & 1 & 0 & 1 & 0 & 0 & 0 & 1 \\ 0 & 0 & 1 & -3 & 1 & 0 & 0 & 0 & 1 & 0 \\ 0 & 1 & 0 & 1 & -3 & 1 & 0 & 0 & 0 & 0 \\ 0 & 0 & 1 & 0 & 1 & -3 & 1 & 0 & 0 & 0 \\ 0 & 1 & 0 & 0 & 0 & 1 & -4 & 1 & 0 & 1 \\ 0 & 0 & 0 & 0 & 0 & 0 & 1 & -2 & 1 & 0 \\ 0 & 0 & 0 & 1 & 0 & 0 & 0 & 1 & -2 & 0 \\ 1 & 0 & 1 & 0 & 0 & 0 & 1 & 0 & 0 & -3 \end{pmatrix}.$$

It is obvious that such complex network is connected. Matrix G is

$$\begin{pmatrix} -1 & \frac{1}{3} & \frac{1}{3} & 0 & 0 & 0 & 0 & 0 & 0 & \frac{1}{3} \\ \frac{1}{4} & -1 & \frac{1}{4} & 0 & \frac{1}{4} & 0 & \frac{1}{4} & 0 & 0 & 0 \\ \frac{1}{5} & \frac{1}{5} & -1 & \frac{1}{5} & 0 & \frac{1}{5} & 0 & 0 & 0 & \frac{1}{5} \\ 0 & 0 & \frac{1}{3} & -1 & \frac{1}{3} & 0 & 0 & 0 & \frac{1}{3} & 0 \\ 0 & \frac{1}{3} & 0 & \frac{1}{3} & -1 & \frac{1}{3} & 0 & 0 & 0 & 0 \\ 0 & 0 & \frac{1}{3} & 0 & \frac{1}{3} & -1 & \frac{1}{3} & 0 & 0 & 0 \\ 0 & \frac{1}{4} & 0 & 0 & 0 & \frac{1}{4} & -1 & \frac{1}{4} & 0 & \frac{1}{4} \\ 0 & 0 & 0 & 0 & 0 & 0 & \frac{1}{2} & -1 & \frac{1}{2} & 0 \\ 0 & 0 & 0 & \frac{1}{2} & 0 & 0 & 0 & \frac{1}{2} & -1 & 0 \\ \frac{1}{3} & 0 & \frac{1}{3} & 0 & 0 & 0 & \frac{1}{3} & 0 & 0 & -1 \end{pmatrix}.$$

We can easily obtain

$$\Lambda = \text{diag}\{0, -0.3764, -0.5757, -0.6968, -0.9427, -1.1635, -1.2721, -1.4823, -1.5626, -1.9279\}.$$

We take $U = I_2$, $H = I_3$, then

$$M = \begin{pmatrix} 0.2 & 0.5 \\ 0.1 & 0.2 \end{pmatrix},$$

$$E = \begin{pmatrix} 0.3 & 0.5 \\ 0.1 & 0.4 \end{pmatrix},$$

$$\hat{f}_1(\hat{x}_i^1(t)) = \begin{pmatrix} -4\hat{x}_{i1}^1 \\ 3(\hat{x}_{i1}^1)^2 - 5\hat{x}_{i2}^1 \end{pmatrix},$$

$\hat{x}_i^1(t) = (\hat{x}_{i1}^1, \hat{x}_{i2}^1)^T \in \mathbb{R}^2, i = 1, 2, \cdots, 10$. It is obvious that $s(t) = (0, 0)^T \in \mathbb{R}^2$ is an equilibrium of the following system and $D\hat{f}_1(s(t)) = \text{diag}(-4, -5)$:

$$\begin{pmatrix} \dot{s}_1(t) \\ \dot{s}_2(t) \end{pmatrix} = \begin{pmatrix} -4s_1(t) \\ 3(s_1(t))^2 - 5s_2(t) \end{pmatrix},$$

where $s(t) = (s_1(t), s_2(t))^T \in \mathbb{R}^2$.

Setting $\tau(t) = 0.3 - 0.3e^{-t}$, then $0 \leqslant \tau(t) \leqslant \tau = 0.3$, $\dot{\tau}(t) = 0.3e^{-t} \leqslant 0.3 < 1$, for $t \geqslant 0$. We can find the following matrices P_i and Q_i satisfying (6.6) with $\varepsilon_i = 0.2$,

$$P_i = \text{diag}(5.9876, 6.3452),$$
$$Q_i = \text{diag}(4.5438, 5.8945).$$

According to the Theorem (6.3), complex dynamical network (6.1) with above given parameters achieves local exponential output synchronization. The simulation results are shown in Fig. 6.1.

Example 6.6 Consider a complex dynamical network, in which each node is a 3-dimensional nonlinear system described by

$$\begin{pmatrix} \dot{x}_1 \\ \dot{x}_2 \\ \dot{x}_3 \end{pmatrix} = \begin{pmatrix} -6x_1 - x_1x_2 \\ x_1^2 - 7x_2 \\ 5x_3 + x_2^2 \end{pmatrix}.$$

Take $a = 0.2, b = 0.1, \beta_\omega = 1$,

$$\Gamma_1 = \begin{pmatrix} 0.1 & 0.3 \\ 0.2 & 0.2 \\ 0.3 & 0.6 \end{pmatrix},$$

$$\Gamma_2 = \begin{pmatrix} 0.2 & 0.9 \\ 0.4 & 0.4 \\ 0.2 & 0.7 \end{pmatrix},$$

$$C = \begin{pmatrix} 1 & 0 & 0 \\ 0 & 1 & 0 \end{pmatrix}.$$

The matrix L is chosen as

$$\begin{pmatrix} -2 & 0 & 0 & 0 & 0 & 0 & 0 & 0 & 1 & 1 \\ 0 & -1 & 1 & 0 & 0 & 0 & 0 & 0 & 0 & 0 \\ 0 & 1 & -2 & 0 & 0 & 0 & 0 & 0 & 0 & 1 \\ 0 & 0 & 0 & -1 & 0 & 0 & 0 & 0 & 0 & 1 \\ 0 & 0 & 0 & 0 & -1 & 0 & 0 & 0 & 0 & 1 \\ 0 & 0 & 0 & 0 & 0 & -1 & 0 & 0 & 1 & 0 \\ 0 & 0 & 0 & 0 & 0 & 0 & -1 & 1 & 0 & 0 \\ 0 & 0 & 0 & 0 & 0 & 0 & 1 & -2 & 1 & 0 \\ 1 & 0 & 0 & 0 & 0 & 1 & 0 & 1 & -4 & 1 \\ 1 & 0 & 1 & 1 & 1 & 0 & 0 & 0 & 1 & -5 \end{pmatrix}.$$

It is obvious that such complex network is connected. We take $U = I_2$, $H = I_3$, then

$$M = \begin{pmatrix} 0.1 & 0.3 \\ 0.2 & 0.2 \end{pmatrix},$$

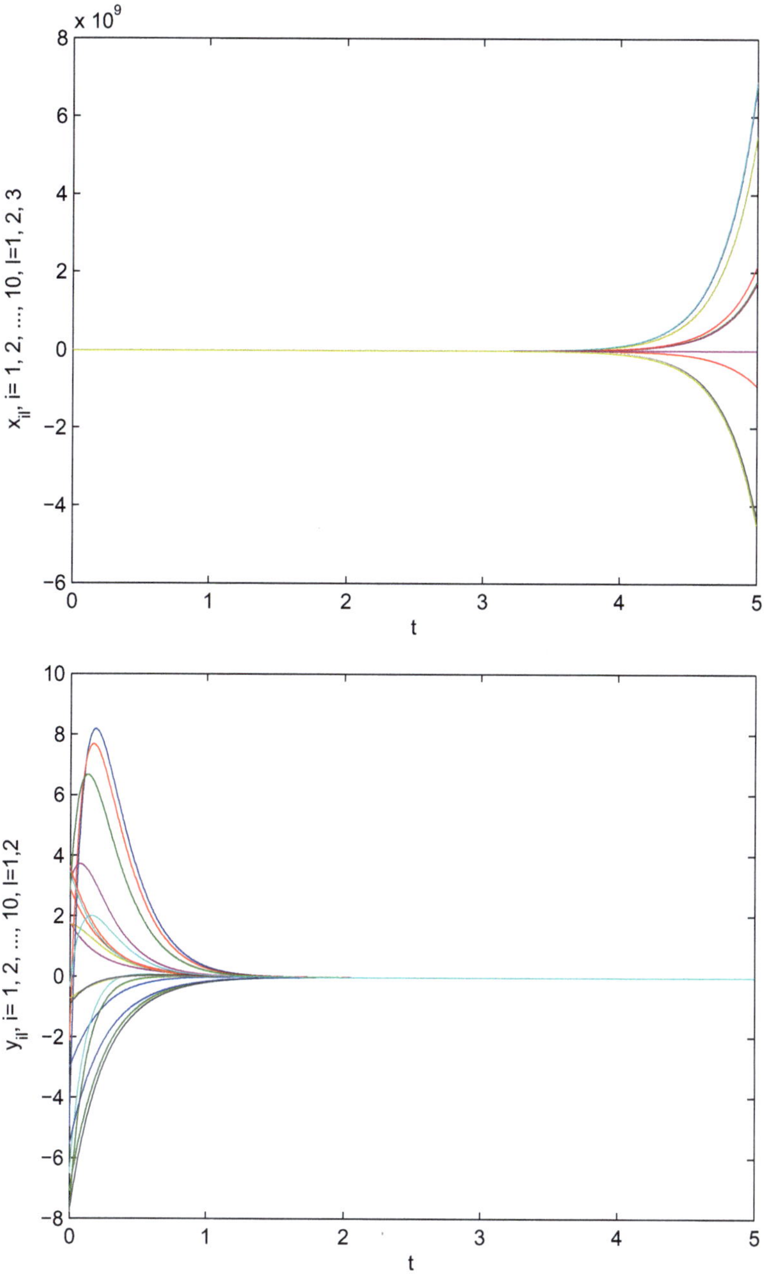

Fig. 6.1 The change process of the state variables and output variables

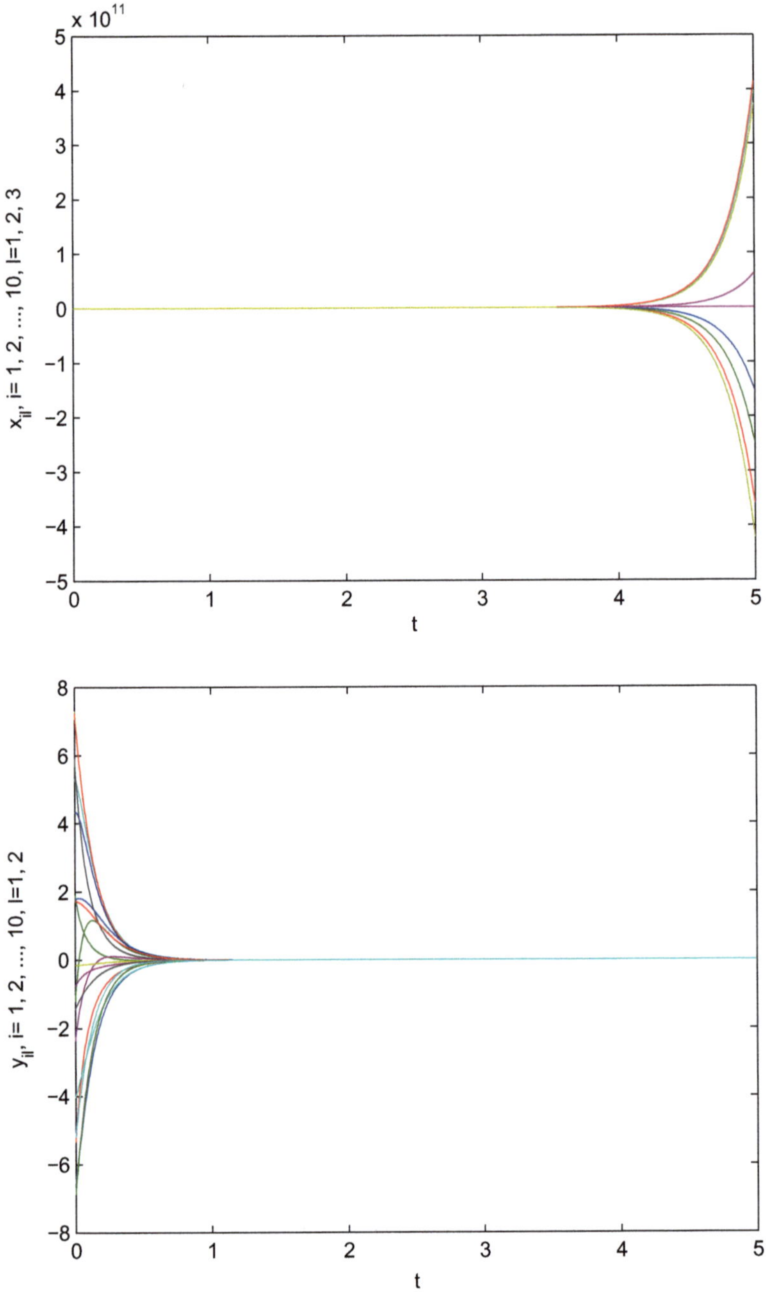

Fig. 6.2 The change process of the state variables and output variables

$$E = \begin{pmatrix} 0.2 \ 0.9 \\ 0.4 \ 0.4 \end{pmatrix},$$

$$\hat{f}_1(\hat{x}_i^1(t)) = \begin{pmatrix} -6\hat{x}_{i1}^1 - \hat{x}_{i1}^1\hat{x}_{i2}^1 \\ (\hat{x}_{i1}^1)^2 - 7\hat{x}_{i2}^1 \end{pmatrix},$$

$$\hat{x}_i^1(t) = (\hat{x}_{i1}^1, \hat{x}_{i2}^1)^T \in \mathbb{R}^2,$$

$i = 1, 2, \cdots, 10$, $s(t) = (0, 0) \in \mathbb{R}^2$ is an equilibrium of the following system

$$\dot{s}(t) = \hat{f}_1(s(t)) = \begin{pmatrix} -6s_1(t) - s_1(t)s_2(t) \\ s_1^2(t) - 7s_2(t) \end{pmatrix},$$

where $s(t) = (s_1(t), s_2(t))^T \in \mathbb{R}^2$. In this case, we can get (6.15) if we take $\eta = 6$, $P = \mathrm{diag}(1, 1)$, $\Delta = \mathrm{diag}(0, 0)$. Setting $\tau(t) = 0.2 - 0.2e^{-t}$, then $0 \leqslant \tau(t) \leqslant \tau = 0.2$, $\dot{\tau}(t) = 0.2e^{-t} \leqslant 0.2 < 1$, for $t \geqslant 0$.

According to the proof of Theorem (6.4), the network (6.1) achieves global exponential output synchronization if (6.13) and (6.15) hold. We can find the matrix $Q = 3.2475I_{20}$ satisfying (6.13) with $\varepsilon = 0.3$. Therefore, complex network (6.1) with above given parameters achieves global exponential output synchronization. The simulation results are shown in Fig. 6.2.

Remark 6.7 According to the change process of the state variables and output variables in Figs. 6.1 and 6.2, we clearly see that the outputs are synchronization, but the states are asynchronous. The main reason is that the outputs are relate to only part states. Therefore, only part states are needed to make the output synchronization to come true.

6.5 Conclusion

A new complex delayed dynamical network model with output coupling has first been introduced. We have studied the output synchronization of the proposed network model. Several local and global exponential output synchronization criteria have been established. Two examples have been given to show the effectiveness of the obtained results.

Chapter 7
Adaptive Output Synchronization of CDDNs with Output Coupling

7.1 Introduction

In the real world, complex networks can be seen everywhere, and have been considered as a fundamental tool to understand dynamical behavior and the response of real systems such as food webs, communication networks, social networks, power grids, cellular networks, World Wide Web, metabolic systems, disease transmission networks, and many others [2, 3]. The topology and dynamical behavior of various complex networks have been extensively studied by researchers [4, 5]. Especially, as one of the most significant and interesting dynamical properties of the complex networks, synchronization has received much of the focus in recent years. So far, a great many important results on synchronization have been obtained for various complex networks such as time invariant, time-varying, and impulsive network models; see [6–21] and relevant references therein.

It should be noticed that the state synchronization of complex networks with state coupling was considered in these articles (see also the above mentioned references). Practically, there are two kinds of coupling forms in complex networks: state coupling and output coupling. As we know, many phenomena in nature can be modeled as complex networks with output coupling [22, 23]. Nevertheless, there are very few works on complex dynamical networks with output coupling [22–24]. To our knowledge, Jiang, Tang and Chen [22] first introduced a complex network model with output coupling. Some conditions for synchronization were established based on the Lyapunov stability theory. In [23], Chen proposed a complex network model with output coupling and random sensor delay. A sufficient synchronization condition was given to ensure that the proposed network model is exponentially mean-square stable. One should note that the state synchronization was investigated in [22, 23]. It is well known that the node state in complex networks is difficult to be observed or measured, even the node state can not be observed or measured at all. Moreover, in many circumstances, only part states are needed to make the synchronization to come true. For these phenomena, it is more interesting

J.-L. Wang et al., *Analysis and Control of Output Synchronization for Complex Dynamical Networks*, https://doi.org/10.1007/978-981-13-1352-3_7

to study the output synchronization of complex networks [24, 25]. For instance, Wang and Wu [25] discussed the output synchronization of a class of impulsive complex dynamical networks with time-varying delay. By constructing suitable Lyapunov functionals, some useful conditions were obtained to guarantee the local and global exponential output synchronization of the impulsive complex networks. Unfortunately, very few authors have considered the output synchronization for complex networks with output coupling [24]. So, it is essential to further study the output synchronization of complex dynamical networks with time-varying delay and output coupling.

Recently, the synchronization control problem of CDNs has become a very hot topic in both theoretical research and practical applications. Among the existing results, some researchers focused on adaptive control [6, 7, 153, 157–163] on complex dynamical networks by applying state feedback controllers. For instance, Zhou, Lu and Lü [6] investigated the locally and globally adaptive synchronization of an uncertain complex dynamical network. In [158], Ji et al. proposed an adaptive control method to achieve the lag synchronization between uncertain complex dynamical network having delayed coupling and a nonidentical reference node. Wang et al. [153] proved that the states of a weighted complex dynamical network with time-varying delay can globally asymptotically synchronize onto a desired orbit under the designed controllers, and the adaptive controllers have strong robustness against asymmetric coupling matrix, time-varying weights, delays, and noise. However, the node state often can not be observed or measured, which makes controller design very difficult. Moreover, these adaptive control strategies [6, 7, 153, 157, 159, 160, 162, 163] are based on a special solution of an isolate node of the networks, which may be difficult to obtain in some engineering applications. To overcome these difficulties, some adaptive output feedback controllers are proposed in this chapter.

Motivated by the above discussions, in this chapter, we propose a complex delayed dynamical network with output coupling. The objective of this chapter is to study the adaptive output synchronization of the proposed network model. By constructing appropriate Lyapunov functionals and utilizing adaptive control technique, some sufficient conditions on output synchronization are derived for the proposed network model.

The rest of this chapter is organized as follows. In Sect. 7.2, our mathematical model of complex delayed dynamical network is presented and some preliminaries are given. The main results of this chapter are given in Sects. 7.3 and 7.4. In Sect. 7.5, two numerical examples are provided to illustrate the effectiveness of the theoretical results. Finally, Sect. 7.6 concludes the investigation.

7.2 Network Model

In this chapter, we consider a complex network consisting of N identical nodes with diffusive and output coupling, in which each node is an n-dimensional dynamical system. The mathematical model of the network can be described as follows:

$$\begin{cases} \dot{x}_i(t) = f(x_i(t)) + a\sum_{j=1}^{N} G_{ij} y_j(t-\tau(t)) + u_i(t), \\ y_i(t) = \qquad\qquad Cx_i(t) \end{cases} \tag{7.1}$$

where $i = 1, 2, \cdots, N$. The function $f(\cdot)$, describing the local dynamics of the nodes, is continuous and capable of producing various rich dynamical behaviors; $x_i(t) = (x_{i1}(t), x_{i2}(t), \cdots, x_{in}(t))^T \in \mathbb{R}^n$ is the state variable of node i; $y_i(t) = (y_{i1}(t), y_{i2}(t), \cdots, y_{in}(t))^T \in \mathbb{R}^n$ is the output vector of node i; $u_i(t) \in \mathbb{R}^n$ is the control input; $\tau(t)$ is the time-varying delay with $0 \leqslant \tau(t) \leqslant \tau$; a is a positive real number, which represents the overall coupling strength; the output matrix $C = \text{diag}(c_1, c_2, \cdots, c_n)$ is a semi-positive definite matrix; $G = (G_{ij})_{N\times N}$ represents the topological structure of network and coupling strength between nodes, where G_{ij} is defined as follows: if there exists a connection between node i and node j, then $G_{ij} = G_{ji} > 0$; otherwise, $G_{ij} = G_{ji} = 0 (i \neq j)$, and the diagonal elements of matrix G are defined by

$$G_{ii} = -\sum_{\substack{j=1 \\ j\neq i}}^{N} G_{ij}, \quad i = 1, 2, \cdots, N.$$

In this chapter, we always assume that complex network (7.1) is connected. Since network (7.1) is connected in the sense of having no isolated clusters, which means that the coupling matrix G is irreducible. The initial condition associated with the complex network (7.1) is given in form

$$x_i(s) = \Phi_i(s),$$
$$y_i(s) = C\Phi_i(s),$$

where $i = 1, 2, \cdots, N$, $\Phi_i(s) \in \mathrm{C}([-\tau, 0], \mathbb{R}^n)$.

Remark 7.1 In [22], Jiang, Tang and Chen first introduced a complex network model with output coupling without time delay. However, time delays always exist in complex networks due to the finite speeds of transmission and/or the traffic congestion, and most of delays is notable. So it is crucial for us to take the delay into the consideration when we study complex networks. Moreover, absolute constant delay may be scarce and delays are frequently varied with time. Therefore, a complex network model with output coupling and time-varying delay is considered in this chapter. In addition, another issue of importance deserving attention is the fact that many existing complex network models with state coupling can be represented by (7.1) through an appropriate choice of the parameters, e.g., see also [164–168].

Remark 7.2 In recent years, a lot of efforts have been made to study the adaptive state synchronization of complex networks with state coupling [153, 157–163]. Unfortunately, the node state in complex networks is difficult to be observed or measured, even the node state can not be observed or measured at all. Moreover, in many circumstances, only part states are needed to make the synchronization to come true. Therefore, it is more interesting to study the output synchronization of complex networks. To the best of our knowledge, this is the first time to consider the adaptive output synchronization of CDDNs with output coupling, which is a very important and interesting problem.

In what follows, we introduce a useful definition.

Definition 7.3 The complex network (7.1) is said to achieve output synchronization if

$$\lim_{t\to+\infty}\left\| y_i(t)-\frac{1}{N}\sum_{j=1}^{N} y_j(t)\right\| = 0,\ \text{for all } i=1,2,...,N.$$

7.3 Adaptive Output Synchronization of Complex Delayed Dynamical Network with Positive Definite Output Matrix

This section discusses the adaptive output synchronization of complex delayed dynamical network (7.1) with positive definite output matrix. Several output synchronization criteria are obtained by constructing appropriate Lyapunov functionals and utilizing the designed adaptive controllers.

For convenient analysis, we let

$$\hat{f}(x(t)) = Cf(C^{-1}x(t))$$

where $x(t)\in\mathbb{R}^n$. Then we can get from (7.1) that

$$\dot{y}_i(t) = \hat{f}(y_i(t)) + a\sum_{j=1}^{N} G_{ij}Cy_j(t-\tau(t)) + Cu_i(t) \tag{7.2}$$

where $i=1,2,\cdots,N$.

In order to obtain our main results, we make the following assumption:

(A1) There exist positive definite diagonal matrix $P=\text{diag}(p_1,p_2,\cdots,p_n)$ and diagonal matrix $\Delta=\text{diag}(\delta_1,\delta_2,\cdots,\delta_n)$ such that $\hat{f}$ satisfies the following inequality:

$$(x-y)^T P[(\hat{f}(x)-\hat{f}(y))-\Delta(x-y)] \leqslant -\eta(x-y)^T(x-y)$$

for some $\eta>0$ and all $x,y\in\mathbb{R}^n$.

7.3.1 Adaptive Output Synchronization of Complex Network with Output Coupling

Let $V = \{1, 2, \cdots, N\}$ and $E \subset V \times V$ respectively denote the set of nodes and the set of undirected links in the network (7.1). $\hat{G}(t) = (\hat{G}_{ij}(t))_{N\times N}$ is a time-varying matrix, where $\hat{G}_{ij}(t)$ is defined as follows: if there is an edge between node i and node j at time t, then the element $\hat{G}_{ij}(t) = \hat{G}_{ji}(t) > 0$; otherwise, $\hat{G}_{ij}(t) = \hat{G}_{ji}(t) = 0 (i \neq j)$, and the diagonal elements of matrix $\hat{G}(t)$ are defined by

$$\hat{G}_{ii}(t) = -\sum_{\substack{j=1\\ j\neq i}}^{N} \hat{G}_{ij}(t), \quad i = 1, 2, \cdots, N.$$

In this chapter, the topological structure of network (7.1) is fixed. That is, if there is no connection between nodes i and j, then $\hat{G}_{ij}(t) = \hat{G}_{ji}(t) = 0$ for all t.

Theorem 7.4 *Let* (A1) *hold and* $\dot{\tau}(t) \leqslant \sigma < 1$. *Then, complex network* (7.1) *with positive definite output matrix achieves output synchronization under the adaptive controllers*

$$u_i(t) = a\sum_{j=1}^{N} \hat{G}_{ij}(t) y_j(t), \quad i = 1, 2, \cdots, N \tag{7.3}$$

and updating laws

$$\begin{aligned} \dot{\hat{G}}_{ij}(t) &= \beta_{ij}(y_i(t) - y_j(t))^T PC(y_i(t) - y_j(t)), \\ \hat{G}_{ij}(0) &= \hat{G}_{ji}(0) > 0, \end{aligned} \tag{7.4}$$

$(i, j) \in E$, *where* $\beta_{ij} = \beta_{ji}$ *are positive constants.*

Proof Let $\bar{y}(t) = \frac{1}{N}\sum_{j=1}^{N} y_j(t)$. Then, we have

$$\dot{\bar{y}}(t) = \frac{1}{N}\sum_{j=1}^{N} \hat{f}(y_j(t)). \tag{7.5}$$

Define $e_i(t) = y_i(t) - \bar{y}(t)$, then the dynamics of the output error vector $e_i(t)$ is governed by the following equation:

$$\dot{e}_i(t) = \hat{f}(y_i(t)) - \frac{1}{N}\sum_{j=1}^{N} \hat{f}(y_j(t)) + a\sum_{j=1}^{N} \hat{G}_{ij}(t) C e_j(t)$$

$$+a\sum_{j=1}^{N} G_{ij}Ce_j(t-\tau(t)), \tag{7.6}$$

where $i = 1, 2, \cdots, N$.

Construct a Lyapunov functional for the system (7.6) as follows:

$$V(t) = \frac{1}{2}\sum_{i=1}^{N} e_i^T(t)Pe_i(t) + \sum_{i=1}^{N}\sum_{\substack{j=1\\ j\neq i}}^{N} \frac{a}{4\beta_{ij}}(\hat{G}_{ij}(t) - b_{ij})^2$$
$$+\frac{a}{2}\sum_{i=1}^{N}\sum_{j=1}^{N} \frac{|G_{ij}|}{1-\sigma}\int_{t-\tau(t)}^{t} e_j^T(s)PCe_j(s)ds, \tag{7.7}$$

where $b_{ij} = b_{ji}$ are nonnegative constants, and $b_{ij} = 0$ if and only if $\hat{G}_{ij}(t) = 0$.

Calculating the time derivative of $V(t)$ along the trajectory of system (7.6), we can get

$$\dot{V}(t) \leqslant \sum_{i=1}^{N} e_i^T(t)P\dot{e}_i(t) + \sum_{i=1}^{N}\sum_{\substack{j=1\\ j\neq i}}^{N} \frac{a}{2\beta_{ij}}(\hat{G}_{ij}(t) - b_{ij})\dot{\hat{G}}_{ij}(t)$$
$$-\frac{a}{2}\sum_{i=1}^{N}\sum_{j=1}^{N} |G_{ij}|e_j^T(t-\tau(t))PCe_j(t-\tau(t))$$
$$+\frac{a}{2}\sum_{i=1}^{N}\sum_{j=1}^{N} \frac{|G_{ij}|}{1-\sigma}e_j^T(t)PCe_j(t)$$
$$= \sum_{i=1}^{N} e_i^T(t)P[\hat{f}(y_i(t)) - \frac{1}{N}\sum_{j=1}^{N} \hat{f}(y_j(t))$$
$$+a\sum_{j=1}^{N} \hat{G}_{ij}(t)Ce_j(t) + a\sum_{j=1}^{N} G_{ij}Ce_j(t-\tau(t))]$$
$$+\frac{a}{2}\sum_{i=1}^{N}\sum_{j=1}^{N} \frac{|G_{ij}|}{1-\sigma}e_j^T(t)PCe_j(t)$$
$$+\frac{a}{2}\sum_{i=1}^{N}\sum_{\substack{j=1\\ j\neq i}}^{N}(\hat{G}_{ij}(t) - b_{ij})(y_i(t) - y_j(t))^T PC(y_i(t) - y_j(t))$$

$$
\begin{aligned}
&-\frac{a}{2}\sum_{i=1}^{N}\sum_{j=1}^{N}|G_{ij}|e_j^T(t-\tau(t))PCe_j(t-\tau(t))\\
&\leqslant \sum_{i=1}^{N}e_i^T(t)P[\hat{f}(y_i(t))-\frac{1}{N}\sum_{j=1}^{N}\hat{f}(y_j(t))-\hat{f}(\bar{y}(t))+\hat{f}(\bar{y}(t))\\
&+a\sum_{j=1}^{N}\hat{G}_{ij}(t)Ce_j(t)]+\frac{a}{2}\sum_{i=1}^{N}\sum_{j=1}^{N}\frac{|G_{ij}|}{1-\sigma}e_j^T(t)PCe_j(t)\\
&+\frac{a}{2}\sum_{i=1}^{N}\sum_{\substack{j=1\\ j\neq i}}^{N}(\hat{G}_{ij}(t)-b_{ij})(e_i(t)-e_j(t))^T PC(e_i(t)-e_j(t))\\
&+\frac{a}{2}\sum_{i=1}^{N}\sum_{j=1}^{N}|G_{ij}|e_i^T(t)PCe_i(t).
\end{aligned}
\tag{7.8}
$$

Since $\sum_{i=1}^{N}e_i^T(t)=0$, we can obtain

$$
\sum_{i=1}^{N}e_i^T(t)P\left(\hat{f}(\bar{y}(t))-\frac{1}{N}\sum_{j=1}^{N}\hat{f}(y_j(t))\right)=0. \tag{7.9}
$$

According to **(A1)**, we have

$$
\sum_{i=1}^{N}e_i^T(t)P\left(\hat{f}(y_i(t))-\hat{f}(\bar{y}(t))\right)\leqslant\sum_{i=1}^{N}e_i^T(t)(-\eta I_n+P\Delta)e_i(t). \tag{7.10}
$$

Define the matrix $B=(b_{ij})_{N\times N}$, where $b_{ii}=-\sum_{\substack{j=1\\ j\neq i}}^{N}b_{ij},\ i=1,2,\cdots,N$. Then, we can get

$$
\begin{aligned}
&\sum_{i=1}^{N}\sum_{\substack{j=1\\ j\neq i}}^{N}(\hat{G}_{ij}(t)-b_{ij})(e_i(t)-e_j(t))^T PC(e_i(t)-e_j(t))\\
&=-2\sum_{i=1}^{N}\sum_{j=1}^{N}(\hat{G}_{ij}(t)-b_{ij})e_i^T(t)PCe_j(t).
\end{aligned}
\tag{7.11}
$$

It follows from (7.8), (7.9), (7.10) and (7.11) that

$$
\begin{aligned}
\dot{V}(t) &\leqslant \sum_{i=1}^{N} e_i^T(t)(-\eta I_n + P\Delta) + \frac{a(2-\sigma)|G_{ii}|}{1-\sigma}PC)e_i(t) \\
&\quad + a\sum_{i=1}^{N}\sum_{j=1}^{N} b_{ij}e_i^T(t)PCe_j(t) \\
&= e^T(t)[-\eta(I_N \otimes I_n) + (I_N \otimes (P\Delta)) \\
&\quad + a(B \otimes (PC)) + \frac{a(2-\sigma)}{1-\sigma}(\bar{G} \otimes (PC))]e(t),
\end{aligned} \tag{7.12}
$$

where $\bar{G} = \text{diag}(|G_{11}|, |G_{22}|, \cdots, |G_{NN}|)$, $e(t) = (e_1^T(t), e_2^T(t), \cdots, e_N^T(t))^T$.

According to Lemma 1.11, there obviously exists a unitary matrix $\phi = (\phi_1, \phi_2, \cdots, \phi_N) \in \mathbb{R}^{N\times N}$ such that $\phi^T B\phi = \Lambda$ with $\Lambda = \text{diag}(\lambda_1, \lambda_2, \cdots, \lambda_N)$. $\lambda_i, i = 1, 2, \cdots, N$, are the eigenvalues of B and $0 = \lambda_1 > \lambda_2 \geqslant \lambda_3 \geqslant \cdots \geqslant \lambda_N$. Let $z(t) = (z_1^T(t), z_2^T(t), \cdots, z_N^T(t))^T = (\phi^T \otimes I_n)e(t)$. Since $\phi_1 = \frac{1}{\sqrt{N}}(1, 1, \cdots, 1)^T$, one has $z_1(t) = (\phi_1^T \otimes I_n)e(t) = 0$. Then, we can derive from (7.12) that

$$
\begin{aligned}
\dot{V}(t) &\leqslant e^T(t)[-\eta(I_N \otimes I_n) + (I_N \otimes (P\Delta)) + a(\phi \otimes I_n)(\Lambda \otimes (PC))(\phi^T \otimes I_n) \\
&\quad + \frac{a(2-\sigma)}{1-\sigma}(\bar{G} \otimes (PC))]e(t) \\
&= e^T(t)[-\eta(I_N \otimes I_n) + (I_N \otimes (P\Delta)) + \frac{a(2-\sigma)}{1-\sigma}(\bar{G} \otimes (PC))]e(t) \\
&\quad + az^T(t)(\Lambda \otimes (PC))z(t).
\end{aligned} \tag{7.13}
$$

Since P and C are positive definite diagonal matrices, we have

$$
z^T(t)(\Lambda \otimes (PC))z(t) \leqslant \lambda_2 z^T(t)(I_N \otimes (PC))z(t).
$$

Therefore,

$$
\begin{aligned}
\dot{V}(t) &\leqslant e^T(t)[-\eta(I_N \otimes I_n) + (I_N \otimes (P\Delta)) + \frac{a(2-\sigma)}{1-\sigma}(\bar{G} \otimes (PC))]e(t) \\
&\quad + a\lambda_2 z^T(t)(I_N \otimes (PC))z(t) \\
&= e^T(t)[-\eta(I_N \otimes I_n) + (I_N \otimes (P\Delta)) + \frac{a(2-\sigma)}{1-\sigma}(\bar{G} \otimes (PC))]e(t) \\
&\quad + a\lambda_2 e^T(t)(\phi \otimes I_n)(I_N \otimes (PC))(\phi^T \otimes I_n)e(t) \\
&= e^T(t)[-\eta(I_N \otimes I_n) + (I_N \otimes (P\Delta)) + \frac{a(2-\sigma)}{1-\sigma}(\bar{G} \otimes (PC)) \\
&\quad + a\lambda_2(I_N \otimes (PC))]e(t).
\end{aligned} \tag{7.14}
$$

By selecting b_{ij} sufficiently large such that

$$a\lambda_2\underline{c} + \frac{a(2-\sigma)\bar{g}\bar{c}}{1-\sigma} + \bar{\delta} \leqslant 0, \tag{7.15}$$

one obtains

$$(I_N \otimes (P\Delta)) + \frac{a(2-\sigma)}{1-\sigma}(\bar{G} \otimes (PC)) + a\lambda_2(I_N \otimes (PC)) \leqslant 0,$$

where

$$\underline{c} = \min_{i=1,2,\cdots,N}\{p_i c_i\},\ \bar{c} = \max_{i=1,2,\cdots,N}\{p_i c_i\},$$

$$\bar{g} = \max_{i=1,2,\cdots,N}\{|G_{ii}|\},\ \bar{\delta} = \max_{i=1,2,\cdots,N}\{p_i\delta_i\}.$$

Therefore,

$$\dot{V}(t) \leqslant -\eta e^T(t)e(t). \tag{7.16}$$

Since $V(t)$ is monotonically decreasing and bounded, we can conclude that $V(t)$ asymptotically converges to a finite non-negative value. From (7.16), we have

$$\|e(t)\|^2 \leqslant -\frac{\dot{V}(t)}{\eta}. \tag{7.17}$$

We can derive from (7.17) that $\lim_{t\to+\infty}\int_0^t \|e(s)\|^2 ds$ exists. Since $0 \leqslant \tau(t) \leqslant \tau$, we can get $\lim_{t\to+\infty}\int_{t-\tau(t)}^t e_j^T(s)PCe_j(s)ds = 0, j = 1, 2, \cdots, N$. On the other hand, since $\hat{G}_{ij}(t)$ is monotonically increasing and bounded [see (7.4)], one can conclude that $\hat{G}_{ij}(t)((i, j) \in E)$ asymptotically converges to a finite positive value. By the definition of $V(t)$, we can derive that $\lim_{t\to+\infty}\sum_{i=1}^N e_i^T(t)Pe_i(t)$ exists and is a non-negative real number. In what follows, we shall prove that

$$\lim_{t\to+\infty}\sum_{i=1}^N e_i^T(t)Pe_i(t) = 0.$$

If this is not true, we have

$$\lim_{t\to+\infty}\sum_{i=1}^N e_i^T(t)Pe_i(t) = \mu > 0.$$

Then, there obviously exists a real number $M > 0$ such that $\sum_{i=1}^{N} e_i^T(t)Pe_i(t) > \frac{\mu}{2}$ for $t \geqslant M$. From (7.16), we can get

$$\dot{V}(t) \leqslant -\eta \frac{\sum_{i=1}^{N} e_i^T(t)Pe_i(t)}{\lambda_M(P)} < -\frac{\eta\mu}{2\lambda_M(P)}, \quad t \geqslant M. \tag{7.18}$$

By integrating (7.18) with respect to t over the time period M to $+\infty$, we can obtain

$$\begin{aligned} -V(M) &\leqslant V(+\infty) - V(M) \\ &= \int_M^{+\infty} \dot{V}(t)dt \\ &< -\int_M^{+\infty} \frac{\eta\mu}{2\lambda_M(P)}dt \\ &= -\infty. \end{aligned}$$

This yields a contradiction, and so $\lim_{t\to+\infty} \sum_{i=1}^{N} e_i^T(t)Pe_i(t) = 0$. Then, we can easily obtain $\lim_{t\to+\infty} \|e(t)\| = 0$. Therefore, the complex network (7.1) achieves output synchronization under the adaptive controllers (7.3) and updating laws (7.4). The proof is completed.

Remark 7.5 In recent years, many adaptive control schemes [6, 7, 153, 157, 159, 160, 162, 163] were proposed to guarantee the synchronization of complex networks. However, these adaptive control strategies are based on a special solution of an isolate node of the networks, which may be difficult to obtain in some engineering applications. In this chapter, we remove this restriction.

7.3.2 *Pinning Adaptive Output Synchronization of Complex Network with Output Coupling*

Based on the outputs of all neighboring nodes, an adaptive control scheme is proposed to guarantee the output synchronization of the complex dynamical network (7.1) in the previous subsection. Here, only a portion of neighboring nodes are needed to design the adaptive controllers.

Let $\hat{E}$ denote a subset of $E \subset V \times V$. $A(t) = (A_{ij}(t))_{N\times N}$ is a time-varying matrix, where $A_{ij}(t)$ is defined as follows:

$$A_{ij}(t) = \begin{cases} A_{ij}(t) = A_{ji}(t) > 0, & \text{if } (i,j) \in \hat{E}, \\ A_{ii}(t) = -\sum_{\substack{j=1 \\ j\neq i}}^{N} A_{ij}(t), & \text{if } i = j, \\ 0, & \text{otherwise.} \end{cases}$$

In this subsection, we assume that complex network (7.1) is connected through the undirected edges $\hat{E}$.

Theorem 7.6 *Let* (A1) *hold and* $\dot{\tau}(t) \leqslant \sigma < 1$. *Then, complex network* (7.1) *with positive definite output matrix achieves output synchronization under the adaptive controllers*

$$u_i(t) = a \sum_{j=1}^{N} A_{ij}(t) y_j(t), \quad i = 1, 2, \cdots, N \tag{7.19}$$

and updating laws

$$\begin{aligned} \dot{A}_{ij}(t) &= \beta_{ij}(y_i(t) - y_j(t))^T PC(y_i(t) - y_j(t)), \\ A_{ij}(0) &= A_{ji}(0) > 0, \end{aligned} \tag{7.20}$$

$(i, j) \in \hat{E}$, *where* $\beta_{ij} = \beta_{ji}$ *are positive constants.*

Proof Let $\bar{y}(t) = \frac{1}{N}\sum_{j=1}^{N} y_j(t)$ and $e_i(t) = y_i(t) - \bar{y}(t)$. Then, the dynamics of the output error vector $e_i(t)$ is governed by the following equation:

$$\begin{aligned} \dot{e}_i(t) = \hat{f}(y_i(t)) - \frac{1}{N}\sum_{j=1}^{N} \hat{f}(y_j(t)) + a \sum_{j=1}^{N} A_{ij}(t) C e_j(t) \\ + a \sum_{j=1}^{N} G_{ij} C e_j(t - \tau(t)), \end{aligned} \tag{7.21}$$

where $i = 1, 2, \cdots, N$.

Define the following Lyapunov functional for the system (7.21):

$$\begin{aligned} V(t) = \frac{1}{2}\sum_{i=1}^{N} e_i^T(t) P e_i(t) + \sum_{i=1}^{N} \sum_{(i,j)\in\hat{E}} \frac{a}{4\beta_{ij}} (A_{ij}(t) - b_{ij})^2 \\ + \frac{a}{2}\sum_{i=1}^{N}\sum_{j=1}^{N} \frac{|G_{ij}|}{1-\sigma} \int_{t-\tau(t)}^{t} e_j^T(s) PC e_j(s) ds, \end{aligned} \tag{7.22}$$

where $b_{ij} = b_{ji}$ are nonnegative constants, and $b_{ij} = 0$ if and only if $A_{ij}(t) = 0$.

Then, following similar arguments as in the proof of Theorem 7.4, we can obtain the desired result immediately.

7.4 Adaptive Output Synchronization of Complex Delayed Dynamical Network with Semi-positive Definite Output Matrix

In this section, we extend the above obtained results to the case when the output matrix C is semi-positive definite.

Without loss of generality, rearrange the order of the elements of node state and output and let the first $r(1 \leqslant r < n)$ diagonal elements of matrix C be positive real numbers. That is, $C = \text{diag}(c_1, c_2, \cdots, c_r, 0, \cdots, 0)$.

For the convenience, we denote

$$\begin{aligned}
\hat{C} &= \text{diag}(c_1, c_2, \cdots, c_r),\\
\bar{C} &= \begin{pmatrix}\hat{C} & 0\end{pmatrix} \in \mathbb{R}^{r\times n},\\
x_i^1(t) &= (x_{i1}(t), x_{i2}(t), \cdots, x_{ir}(t))^T \in \mathbb{R}^r,\\
x_i^2(t) &= (x_{i(r+1)}(t), x_{i(r+2)}(t), \cdots, x_{in}(t))^T \in \mathbb{R}^{n-r},\\
x_i(t) &= ((x_i^1(t))^T, (x_i^2(t))^T)^T,\\
y_i^1(t) &= \hat{C}x_i^1(t),\\
y_i(t) &= ((y_i^1(t))^T, 0, 0, \cdots, 0)^T.
\end{aligned}$$

Then, we can get from (7.1) that

$$\dot{y}_i^1(t) = \bar{C}f(x_i(t)) + a\sum_{j=1}^{N} G_{ij}\bar{C}y_j(t-\tau(t)) + \bar{C}u_i(t), \tag{7.23}$$

where $i = 1, 2, \cdots, N$.

In order to obtain our main results, an assumption is introduced.

(A2) Assume that

$$f(x_i(t)) = \begin{pmatrix} f_1(x_i^1(t)) \\ f_2(x_i(t)) \end{pmatrix}$$

where $f_1(x_i^1(t)) \in \mathbb{R}^r$, $f_2(x_i(t)) \in \mathbb{R}^{n-r}$.

According to (7.23) and (A2), we can obtain

$$\dot{y}_i^1(t) = \hat{f}_1(y_i^1(t)) + a\sum_{j=1}^{N} G_{ij}\hat{C}y_j^1(t-\tau(t)) + \bar{C}u_i(t), \tag{7.24}$$

where $i = 1, 2, \cdots, N$, $\hat{f}_1(y_i^1(t)) = \hat{C}f_1(\hat{C}^{-1}y_i^1(t))$.

Next, we make the following assumption:

(A3) There exist positive definite diagonal matrix $Z = \mathrm{diag}(z_1, z_2, \cdots, z_r)$ and diagonal matrix $Q = \mathrm{diag}(q_1, q_2, \cdots, q_r)$ such that $\hat{f}_1$ satisfies the following inequality:

$$(x-y)^T Z[\hat{f}_1(x) - \hat{f}_1(y) - Q(x-y)] \leqslant -\theta(x-y)^T(x-y)$$

for some $\theta > 0$ and all $x, y \in \mathbb{R}^r$.

7.4.1 *Adaptive Output Synchronization of Complex Network with Output Coupling*

For convenient analysis, we let

$$\hat{Z} = \mathrm{diag}(z_1, z_2, \cdots, z_r, 0, \cdots, 0) \in \mathbb{R}^{n\times n}.$$

Theorem 7.7 *Let* (A2) *and* (A3) *hold,* $\dot{\tau}(t) \leqslant \sigma < 1$. *Then, complex network* (7.1) *achieves output synchronization under the adaptive controllers*

$$u_i(t) = a\sum_{j=1}^{N} \hat{G}_{ij}(t)y_j(t), \quad i = 1, 2, \cdots, N \tag{7.25}$$

and updating laws

$$\begin{aligned}\dot{\hat{G}}_{ij}(t) &= \beta_{ij}(y_i(t) - y_j(t))^T \hat{Z}C(y_i(t) - y_j(t)),\\ \hat{G}_{ij}(0) &= \hat{G}_{ji}(0) > 0,\end{aligned} \tag{7.26}$$

$(i, j) \in E$, *where* $\beta_{ij} = \beta_{ji}$ *are positive constants. Here* E *and* $\hat{G}_{ij}(t)$ *have the same meaning as in Sect.* 7.3.1.

Proof Firstly, we can derive from (7.24) and (7.25) that

$$\dot{y}_i^1(t) = \hat{f}_1(y_i^1(t)) + a\sum_{j=1}^{N} \hat{G}_{ij}(t)\hat{C}y_j^1(t) + a\sum_{j=1}^{N} G_{ij}\hat{C}y_j^1(t-\tau(t)), \tag{7.27}$$

where $i = 1, 2, \cdots, N$.

Then, by a minor modification of the proof of Theorem 7.4, we can easily get the conclusion.

7.4.2 *Pinning Adaptive Output Synchronization of Complex Network with Output Coupling*

In this subsection, $\hat{E}$ and $A_{ij}(t)$ denote the same meaning as in Sect. 7.3.2. In addition, we also assume that complex network (7.1) is connected through the undirected edges $\hat{E}$.

Theorem 7.8 *Let* (A2) *and* (A3) *hold,* $\dot{\tau}(t) \leqslant \sigma < 1$. *Then, complex network* (7.1) *achieves output synchronization under the adaptive controllers*

$$u_i(t) = a \sum_{j=1}^{N} A_{ij}(t) y_j(t), \quad i = 1, 2, \cdots, N \tag{7.28}$$

and updating laws

$$\begin{aligned} \dot{A}_{ij}(t) &= \beta_{ij}(y_i(t) - y_j(t))^T \hat{Z} C(y_i(t) - y_j(t)), \\ A_{ij}(0) &= A_{ji}(0) > 0, \quad (i, j) \in \hat{E}, \end{aligned} \tag{7.29}$$

where $\beta_{ij} = \beta_{ji}$ *are positive constants,* $\hat{Z} = \mathrm{diag}(z_1, z_2, \cdots, z_r, 0, \cdots, 0) \in \mathbb{R}^{n \times n}$.

Proof By the similar proof of Theorem 7.6, we can obtain the conclusion. Here we omit its proof to avoid the repetition.

Remark 7.9 In this chapter, some simple adaptive control schemes are proposed to synchronize the outputs of complex delayed dynamical network with output coupling. Here, only node output is needed to design the adaptive controllers. The designed controllers are totally different from some existing ones [6, 7, 153, 157–163]. Obviously, compared with these existing results, our sufficient conditions for network synchronization are very simple and the controllers can be realized more easily.

7.5 Numerical Examples

In this section, we give two examples and their simulation to show the effectiveness of the above obtained theoretical results.

Example 7.10 Consider a complex dynamical network consisting of five identical Chua's circuits, in which each node equation is [18]

$$\begin{pmatrix} \dot{x}_{i1} \\ \dot{x}_{i2} \\ \dot{x}_{i3} \end{pmatrix} = \begin{pmatrix} 10(-x_{i1} + x_{i2} - g(x_{i1})) \\ x_{i1} - x_{i2} + x_{i3} \\ -14.87x_{i2} \end{pmatrix}$$

where $i = 1, 2, \cdots, 5$, $g(x_{i1}) = -0.68x_{i1} + 0.5(-1.27 + 0.68)(|x_{i1} + 1| - |x_{i1} - 1|)$. Take $a = 0.3$, $\tau(t) = 0.4 - 0.4e^{-t}$, $C = \text{diag}(0.5, 0.5, 0.5)$. Then, we have $\dot{\tau}(t) \leqslant 0.4 < 1$ and

$$\hat{f}(x) = \begin{pmatrix} 10(-x_1 + x_2 - \hat{g}(x_1)) \\ x_1 - x_2 + x_3 \\ -14.87x_2 \end{pmatrix}$$

where $x = (x_1, x_2, x_3)^T \in \mathbb{R}^3$, $\hat{g}(x_1) = -0.68x_1 + 0.5(-1.27 + 0.68)(|x_1 + 0.5| - |x_1 - 0.5|)$. It is easy to verify that (A1) is satisfied with $P = I_3$, $\Delta = 11I_3$ and $\eta = 1$.The matrix G is chosen as

$$G = \begin{pmatrix} -1.5 & 0.5 & 0.4 & 0 & 0.6 \\ 0.5 & -0.9 & 0.4 & 0 & 0 \\ 0.4 & 0.4 & -1.5 & 0.7 & 0 \\ 0 & 0 & 0.7 & -1.2 & 0.5 \\ 0.6 & 0 & 0 & 0.5 & -1.1 \end{pmatrix}.$$

Case 1: Choose $\beta_{12} = \beta_{21} = 0.3$, $\beta_{13} = \beta_{31} = 0.4$, $\beta_{15} = \beta_{51} = 0.2$, $\beta_{23} = \beta_{32} = 0.3$, $\beta_{34} = \beta_{43} = 0.4$, $\beta_{45} = \beta_{54} = 0.5$. By using the proposed adaptive controllers (7.3) and updating laws (7.4) for five nodes, the simulation results are shown in Fig. 7.1.

Case 2: Choose $\beta_{12} = \beta_{21} = 0.3$, $\beta_{13} = \beta_{31} = 0.4$, $\beta_{15} = \beta_{51} = 0.2$, $\beta_{34} = \beta_{43} = 0.4$. Under the adaptive controllers (7.19) and updating laws (7.20), the evolutions of the output variables of five nodes are shown in Fig. 7.2.

Next, we analyze the adaptive output synchronization of complex delayed dynamical network with semi-positive definite output matrix.

Example 7.11 Consider a complex dynamical network consisting of five identical nodes, in which each node is a 4-dimensional linear system described by

$$\begin{pmatrix} \dot{v}_{i1} \\ \dot{v}_{i2} \\ \dot{x}_{i1} \\ \dot{x}_{i2} \end{pmatrix} = \begin{pmatrix} 0 \\ 0 \\ v_{i1} \\ v_{i2} \end{pmatrix}$$

where $i = 1, 2, \cdots, 5$. Take $a = 0.4$, $\tau(t) = 0.3 - 0.3e^{-t}$, $C = \text{diag}(1, 1, 0, 0)$. Then we have $\dot{\tau}(t) \leqslant 0.3 < 1$ and $\hat{f}_1(v_i(t)) = 0$, $v_i(t) = (v_{i1}, v_{i2})^T$. It is easy to verify that (A2) and (A3) are satisfied with $Z = Q = I_2$ and $\theta = 1$. The matrix G is chosen as

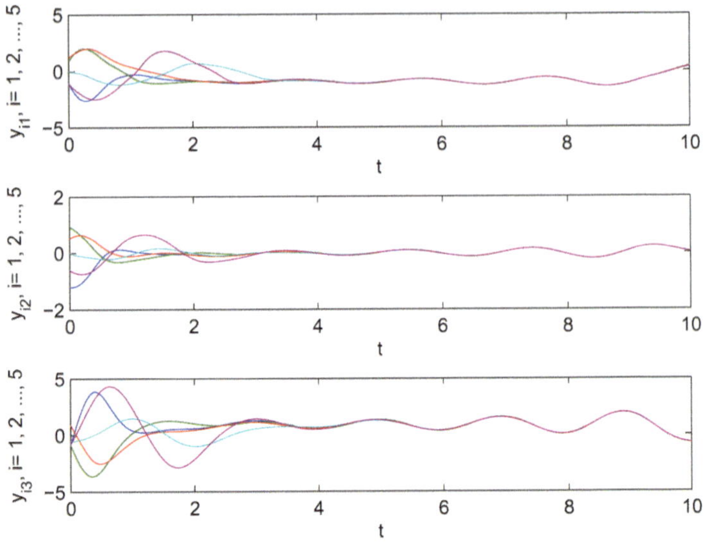

Fig. 7.1 The change processes of the output variables of complex dynamical network (7.1) in time interval [0, 10]

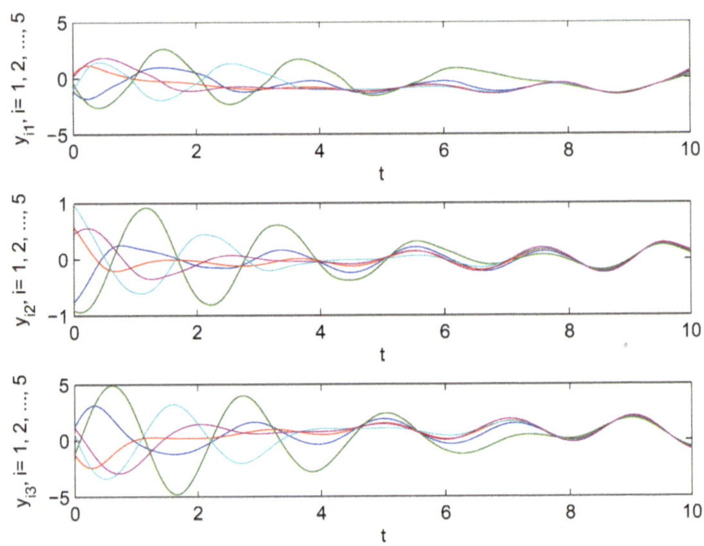

Fig. 7.2 The change processes of the output variables of complex dynamical network (7.1) in time interval [0, 10]

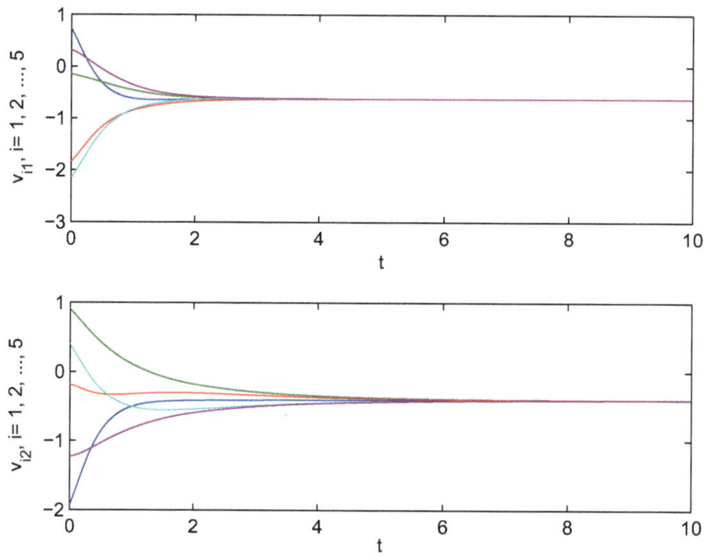

Fig. 7.3 The change processes of $v_{i1}(t)$ and $v_{i2}(t)$, $i = 1, 2, \cdots, 5$

$$G = \begin{pmatrix} -1.7 & 0.3 & 0.4 & 0.4 & 0.6 \\ 0.3 & -0.7 & 0.4 & 0 & 0 \\ 0.4 & 0.4 & -0.8 & 0 & 0 \\ 0.4 & 0 & 0 & -0.7 & 0.3 \\ 0.6 & 0 & 0 & 0.3 & -0.9 \end{pmatrix}.$$

Case 1: Choose $\beta_{12} = \beta_{21} = 0.2$, $\beta_{13} = \beta_{31} = 0.5$, $\beta_{14} = \beta_{41} = 0.3$, $\beta_{15} = \beta_{51} = 0.4$, $\beta_{23} = \beta_{32} = 0.2$, $\beta_{45} = \beta_{54} = 0.3$. With adaptive controllers (7.25) and updating laws (7.26), the evolutions of the output variables of five nodes are shown in Fig. 7.3.

Case 2: Choose $\beta_{12} = \beta_{21} = 0.2$, $\beta_{15} = \beta_{51} = 0.4$, $\beta_{23} = \beta_{32} = 0.2$, $\beta_{45} = \beta_{54} = 0.3$. By using the proposed adaptive controllers (7.28) and updating laws (7.29) for five nodes, the simulation results are shown in Fig. 7.4.

7.6 Conclusion

In this chapter, a new complex delayed dynamical network with output coupling and semi-positive definite output matrix has been introduced. The adaptive output synchronization of the proposed network model has been investigated, and some sufficient conditions have been obtained to guarantee output synchronization of the complex dynamical network. Illustrative simulations have been provided to verify

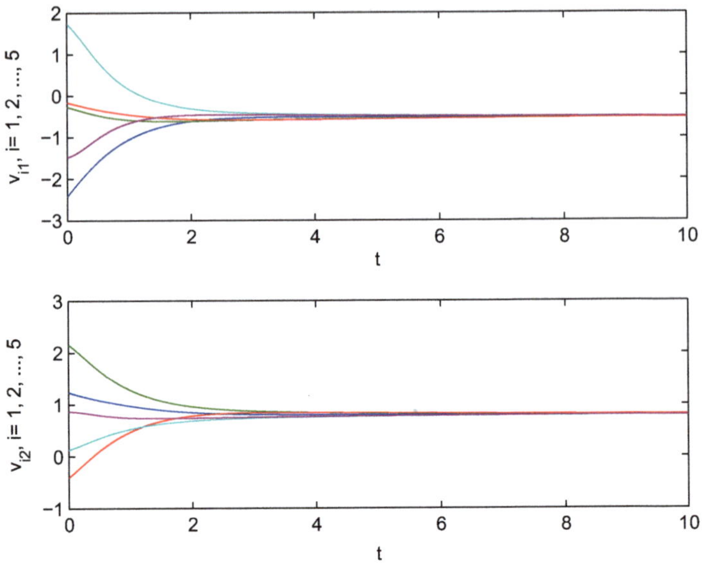

Fig. 7.4 The change processes of $v_{i1}(t)$ and $v_{i2}(t)$, $i = 1, 2, \cdots, 5$

the correctness and effectiveness of the obtained results. In future work, we shall study the output synchronization of complex delayed dynamical networks with output coupling and directed topologies under adaptive control.

Chapter 8
Pinning Synchronization of CDNs with Multi-weights

8.1 Introduction

Recently, the investigation of complex networks synchronization has attracted great interests, and varieties of synchronization criteria have been obtained for different CDNs [88, 92, 169–181]. In [170], the authors considered the finite-time synchronization problem for complex dynamical network with non-derivative and derivative coupling. Xu et al. [172] investigated Markovian jump complex networks with generally uncertain transition rates. By means of Kronecker product and Lyapunov functional method, a sufficient condition was derived to ensure the exponential synchronization for proposed network model. In [176], several criteria for static multisynchronization and dynamical multisynchronization of the coupled multistable neural networks with fixed and switching topologies were obtained by utilizing the impulsive control strategy. Nevertheless, in most existing works on the CDNs [169–176], they always suppose that the node state is a function of time. Actually, the node state is not only relied on the time, but also intensively relied on space variable in many situations [88, 92, 177–181]. Wang et al. [88] proposed two spatially and temporally complex network models. In view of Lyapunov functionals and several inequality techniques, a few sufficient conditions were derived for achieving synchronization by utilizing the designed adaptive laws. In [177], the author analyzed synchronization for coupled reaction-diffusion neural networks with unbounded time delays.

However, most of existing results on synchronization are all based on network models with single weight for not only the ordinary differential equation network models [171, 173, 175, 182–186] but also the partial differential equation network models [92, 177, 179–181, 187]. As is well known, many real-world networks can be described by complex dynamical network models with multi-weights, where nodes are coupled by multiple coupling forms, such as transportation networks, social networks, communication networks, and so on. For instance, individuals or organizations can get in touch with each other by e-mail, mobile phone, facebook,

J.-L. Wang et al., *Analysis and Control of Output Synchronization for Complex Dynamical Networks*, https://doi.org/10.1007/978-981-13-1352-3_8

etc. Considering that every way of communication has different weights, social networks should be modeled by multi-weighted complex networks. Furthermore, in the public traffic roads networks, taking every bus line as the network node, the edge means that two different bus lines have the same bus stops. In this case, there obviously exist different coupling weights in public traffic roads networks, such as coefficient of bus line length, passenger flow density, departing frequency, and so on [39]. Consequently, it is of great importance to investigate multi-weighted CDNs. However, very few researchers have studied synchronization problem of CDNs with multi-weights [39–41]. In [39], the authors investigated the global synchronization of multi-weighted complex networks by changing transfers coefficient, congestion degrees and passenger flow density between different bus lines. An et al. [40] derived a criterion for global synchronization of the public traffic roads networks with multi-weights based on the Lyapunov stability theory. In [41], the authors considered a multi-weighted complex network with time-varying delays and uncertain inner coupling matrices, and a synchronization criterion was established for such network model by exploiting the designed adaptive feedback controller.

Furthermore, because complex networks themselves are usually not synchronized, we must adopt some suitable control strategies to ensure the synchronization in CDNs. From practical perspective, it is very difficult to add control actions to all nodes in a large-scale network. Therefore, many researchers have studied the pinning synchronization problem, and there has been much literature about pinning synchronization of single weighted complex networks [9, 103, 188–196]. Based on the Lyapunov stability theory and graph theory, Li and Yang [191] discussed the synchronization of uncertain complex networks with communication constraints by using a pinning control scheme. In [103], the synchronization for directed and undirected CDNs was investigated via pinning control strategy. Wang et al. [196] studied cluster synchronization of fractional-order CDNs by exploiting pinning adaptive control method. To the best of our knowledge, the pinning synchronization problem of multi-weighted complex networks has not yet been considered.

Motivated by the mentioned discussions, we shall study the pinning synchronization for directed and undirected CDNs with multi-weights. The main contributions in this chapter are given as follows. First, this chapter introduces two multi-weighted complex network models, and these models have several different coupling forms. Second, by pinning a small fraction of nodes with negative feedback controllers, we investigate the synchronization problem of CDNs with multi-weights. In view of Lyapunov functionals and a few inequality techniques, several synchronization criteria are established. Third, some adaptive strategies are designed to derive the suitable coupling strengths for achieving synchronization. Fourth, we come up with some adaptive control schemes to make the feedback gains as small as possible, and two sufficient conditions for synchronization are obtained by utilizing the designed pinning adaptive controllers.

8.2 Pinning Synchronization of CDNs with Multi-weights

8.2.1 Pinning Synchronization for Directed Complex Network with Multi-weights

8.2.1.1 Model Description

The complex network with multi-weights and directed topology is given by

$$\dot{x}_i(t) = f(x_i(t)) + c_1 \sum_{j=1}^{N} G^1_{ij} \Gamma_1 x_j(t) + c_2 \sum_{j=1}^{N} G^2_{ij} \Gamma_2 x_j(t)$$

$$+ \cdots + c_\tau \sum_{j=1}^{N} G^\tau_{ij} \Gamma_\tau x_j(t), \quad i = 1, 2, \cdots, N, \tag{8.1}$$

where $x_i(t) = (x_{i1}(t), x_{i2}(t), \cdots, x_{in}(t))^T \in \mathbb{R}^n$ refers to the state vector of the ith node; $f(x_i(t)) = (f_1(x_{i1}(t)), f_2(x_{i2}(t)), \cdots, f_n(x_{in}(t)))^T : \mathbb{R}^n \to \mathbb{R}^n$ is a continuously differentiable vector function; c_r $(r = 1, 2, \cdots, \tau)$ is a positive real number, which denotes the coupling strength of the rth coupling form; $\Gamma_r \in \mathbb{R}^{n\times n} > 0$ $(r = 1, 2, \cdots, \tau)$ is the inner coupling matrix for the rth coupling form; $G^r = (G^r_{ij})_{N\times N}$ $(r = 1, 2, \cdots, \tau)$ is the coupling configuration matrix representing coupling weights in the rth coupling form, where G^r_{ij} is defined as follows: if there exists a connection from node i to node j, then $G^r_{ij} > 0$; otherwise, $G^r_{ij} = 0 (i \neq j)$; and the diagonal elements of matrix G^r satisfy

$$G^r_{ii} = -\sum_{\substack{j=1 \\ j\neq i}}^{N} G^r_{ij}, i = 1, 2, \cdots, N.$$

Remark 8.1 As far as we know, many real-world networks should be described by multi-weighted complex dynamical network models, where nodes are coupled by multiple coupling forms, such as transportation networks, communication networks, social networks, and so on. Unfortunately, very few researchers have considered the synchronization problem of CDNs with multi-weights [39–41]. Hence, it is very significant to further investigate the synchronization of multi-weighted complex networks. In addition, we always assume that they have the same topology structure for different coupling forms in this section. In other words, $G^r_{ij} = 0$ $(r \in \{1, 2, \cdots, \tau\})$ if and only if $G^\varsigma_{ij} = 0$ for $\varsigma = 1, 2, \cdots, r-1, r+1, \cdots, \tau$.

Suppose that $\hat{s}(t) = (\hat{s}_1(t), \hat{s}_2(t), \cdots, \hat{s}_n(t))^T$ is an arbitrary solution of an isolated node of network (8.1). That is,

$$\dot{\hat{s}}(t) = f(\hat{s}(t)), \quad \hat{s}(t) \in \mathbb{R}^n.$$

In the present work, the function $f_i(\cdot)$ $(i = 1, 2, \cdots, n)$ satisfies Lipschitz condition; that is, there exists positive real number ρ_i such that

$$|f_i(\vartheta_1) - f_i(\vartheta_2)| \leqslant \rho_i |\vartheta_1 - \vartheta_2|$$

for any $\vartheta_1, \vartheta_2 \in \mathbb{R}$.

Definition 8.2 The complex network (8.1) is synchronized if

$$\lim_{t \to +\infty} \|x_i(t) - \hat{s}(t)\| = 0 \text{ for all } i = 1, 2, \cdots, N.$$

Without loss of generality, we suppose that G^r is in the Frobenius normal form. That is,

$$G^r = \begin{pmatrix} \bar{G}_1^r & \bar{G}_{12}^r & \cdots & \bar{G}_{1m}^r \\ 0 & \bar{G}_2^r & \cdots & \bar{G}_{2m}^r \\ \vdots & \vdots & \ddots & \vdots \\ 0 & 0 & \cdots & \bar{G}_m^r \end{pmatrix}, \quad r = 1, 2, \cdots, \tau, \tag{8.2}$$

where $\bar{G}_1^r \in \mathbb{R}^{p_1 \times p_1}, \bar{G}_2^r \in \mathbb{R}^{p_2 \times p_2}, \cdots, \bar{G}_m^r \in \mathbb{R}^{p_m \times p_m}$ are square irreducible matrices.

In order to achieve synchronization, pinning strategy can be applied to a small fraction of nodes in network (8.1). Assume that the nodes $q_1, q_2, \cdots, q_l$ $(1 \leqslant l < N)$ are chosen to be pinned. After that, the pinning controlled network can be written as follows:

$$\begin{aligned} \dot{x}_{q_i}(t) = {} & f(x_{q_i}(t)) + c_1 \sum_{j=1}^{N} G_{q_i j}^1 \Gamma_1 x_j(t) + c_2 \sum_{j=1}^{N} G_{q_i j}^2 \Gamma_2 x_j(t) \\ & + \cdots + c_\tau \sum_{j=1}^{N} G_{q_i j}^\tau \Gamma_\tau x_j(t) + u_{q_i}(t), \\ & i = 1, 2, \cdots, l, \\ \dot{x}_{q_i}(t) = {} & f(x_{q_i}(t)) + c_1 \sum_{j=1}^{N} G_{q_i j}^1 \Gamma_1 x_j(t) + c_2 \sum_{j=1}^{N} G_{q_i j}^2 \Gamma_2 x_j(t) \\ & + \cdots + c_\tau \sum_{j=1}^{N} G_{q_i j}^\tau \Gamma_\tau x_j(t), \\ & i = l+1, l+2, \cdots, N, \end{aligned} \tag{8.3}$$

where

$$u_{q_i}(t) = -\sum_{r=1}^{\tau} c_r h_i^r \Gamma_r (x_{q_i}(t) - \hat{s}(t)), \quad i=1,2,\cdots,l, \tag{8.4}$$

are n-dimensional linear feedback controllers with all the control gains $\mathbb{R} \ni h_i^r > 0$.

From (8.3) and (8.4), one obtains

$$\dot{x}_i(t) = f(x_i(t)) + \sum_{r=1}^{\tau}\sum_{j=1}^{N} c_r G_{ij}^r \Gamma_r x_j(t) - \sum_{r=1}^{\tau} c_r \hat{h}_i^r \Gamma_r (x_i(t) - \hat{s}(t)), \tag{8.5}$$

where $i=1,2,\cdots,N$,

$$\begin{aligned} H^r &= \mathrm{diag}(\hat{h}_1^r, \hat{h}_2^r, \cdots, \hat{h}_N^r) \\ &= \mathrm{diag}(0,\cdots,0,\underbrace{h_1^r}_{q_1},0,\cdots,0,\underbrace{h_2^r}_{q_2},0,\cdots,0,\underbrace{h_l^r}_{q_l},\cdots), \quad r=1,2,\cdots,\tau. \end{aligned}$$

Let $e_i(t) = (e_{i1}(t), e_{i2}(t), \cdots, e_{in}(t))^T = x_i(t) - \hat{s}(t)$. Then, we have

$$\dot{e}_i(t) = f(x_i(t)) - f(\hat{s}(t)) + \sum_{r=1}^{\tau}\sum_{j=1}^{N} c_r \hat{G}_{ij}^r \Gamma_r e_j(t), \tag{8.6}$$

where $\hat{G}^r = (\hat{G}_{ij}^r)_{N\times N} = G^r - H^r, i = 1,2,\cdots,N$.

For the convenience, we denote

$$\Theta = \mathrm{diag}(\rho_1^2, \rho_2^2, \cdots, \rho_n^2),$$
$$\xi = \frac{\lambda_M(I_n + \Theta)}{2\sum_{r=1}^{\tau}\lambda_m(\Gamma_r)}.$$

8.2.1.2 Pinning Synchronization Criteria

Let $H^r = \mathrm{diag}(H_1^r, H_2^r, \cdots, H_m^r)$, $H_i^r \in \mathbb{R}^{p_i\times p_i}, i = 1,2,\cdots,m$. Because $\hat{G}^r = G^r - H^r$ and (8.2), one has

$$\hat{G}^r = \begin{pmatrix} \tilde{G}_1^r & \bar{G}_{12}^r & \cdots & \bar{G}_{1m}^r \\ 0 & \tilde{G}_2^r & \cdots & \bar{G}_{2m}^r \\ \vdots & \vdots & \ddots & \vdots \\ 0 & 0 & \cdots & \tilde{G}_m^r \end{pmatrix}, \quad r=1,2,\cdots,\tau, \tag{8.7}$$

where $\tilde{G}_1^r = \bar{G}_1^r - H_1^r, \tilde{G}_2^r = \bar{G}_2^r - H_2^r, \cdots, \tilde{G}_m^r = \bar{G}_m^r - H_m^r$ are square irreducible matrices.

Obviously, the system (8.6) can be rewritten as

$$\dot{\hat{e}}_i(t) = \hat{f}_i(x(t)) - \hat{f}_i(s^*(t)) + \sum_{r=1}^{\tau} \sum_{j=i+1}^{m} c_r(\bar{G}_{ij}^r \otimes \Gamma_r)\hat{e}_j(t) + \sum_{r=1}^{\tau} c_r(\tilde{G}_i^r \otimes \Gamma_r)\hat{e}_i(t), \quad i = 1, 2, \cdots, m, \tag{8.8}$$

where

$$\begin{aligned}
\hat{e}_i(t) &= (e^T_{p_1+\cdots+p_{i-1}+1}(t),\ e^T_{p_1+\cdots+p_{i-1}+2}(t),\ \cdots, e^T_{p_1+\cdots+p_{i-1}+p_i}(t))^T,\\
s^*(t) &= (\hat{s}^T(t), \hat{s}^T(t), \cdots, \hat{s}^T(t))^T \in \mathbb{R}^{Nn},\\
\hat{f}_i(s^*(t)) &= (\underbrace{f^T(\hat{s}(t)), f^T(\hat{s}(t)), \cdots, f^T(\hat{s}(t))}_{p_i})^T,\\
x(t) &= (x_1^T(t), x_2^T(t), \cdots, x_N^T(t))^T,\\
\hat{f}_i(x(t)) &= (f^T(x_{p_1+\cdots+p_{i-1}+1}(t)), f^T(x_{p_1+\cdots+p_{i-1}+2}(t)), \cdots,\\
&\quad f^T(x_{p_1+\cdots+p_{i-1}+p_i}(t)))^T.
\end{aligned}$$

Theorem 8.3 *If there exist matrices* $0 < \Psi_i = \mathrm{diag}(\Psi_{i1}, \Psi_{i2}, \cdots, \Psi_{ip_i}) \in \mathbb{R}^{p_i \times p_i}$, $i = 1, 2, \cdots, m$, *and positive real numbers* α_i, $i = 1, 2, \cdots, m$, *such that*

$$(\boldsymbol{\alpha}\Psi)\otimes W + \sum_{r=1}^{\tau} c_r\Big(\boldsymbol{\alpha}\Psi G^r + (G^r)^T\Psi\boldsymbol{\alpha} - 2\boldsymbol{\alpha}\Psi H^r\Big)\otimes\Gamma_r < 0 \tag{8.9}$$

where $\boldsymbol{\alpha} = \mathrm{diag}(\alpha_1 I_{p_1}, \alpha_2 I_{p_2}, \cdots, \alpha_m I_{p_m})$, $\Psi = \mathrm{diag}(\Psi_1, \Psi_2, \cdots, \Psi_m)$, $W = I_n + \Theta$, *then the network* (8.3) *is synchronized.*

Proof Firstly, define the following Lyapunov functional for the system (8.8):

$$V_1(t) = \sum_{i=1}^{m} \alpha_i \hat{e}_i^T(t)(\Psi_i \otimes I_n)\hat{e}_i(t). \tag{8.10}$$

Then, one has

$$\begin{aligned}
\dot{V}_1(t) = 2\sum_{i=1}^{m} \alpha_i \hat{e}_i^T(t)\Big[&(\Psi_i \otimes I_n)\big(\hat{f}_i(x(t)) - \hat{f}_i(s^*(t))\big)\\
&+ \sum_{r=1}^{\tau}\sum_{j=i+1}^{m} c_r((\Psi_i \bar{G}_{ij}^r) \otimes \Gamma_r)\hat{e}_j(t)\\
&+ \sum_{r=1}^{\tau} c_r((\Psi_i \tilde{G}_i^r) \otimes \Gamma_r)\hat{e}_i(t)\Big].
\end{aligned} \tag{8.11}$$

Furthermore,

$$
\begin{aligned}
& 2\hat{e}_i^T(t)(\Psi_i \otimes I_n)\big(\hat{f}_i(x(t)) - \hat{f}_i(s^*(t))\big) \\
\leqslant & \sum_{j=1}^{p_i} \Psi_{ij} e^T_{p_1+p_2+\cdots+p_{i-1}+j}(t) e_{p_1+p_2+\cdots+p_{i-1}+j}(t) \\
& + \sum_{j=1}^{p_i} \Psi_{ij} e^T_{p_1+p_2+\cdots+p_{i-1}+j}(t) \Theta e_{p_1+p_2+\cdots+p_{i-1}+j}(t) \\
= & \hat{e}_i^T(t)\left[\Psi_i \otimes (I_n + \Theta)\right] \hat{e}_i(t),
\end{aligned}
$$

where $i = 1, 2, \cdots, m$. Thus,

$$
\begin{aligned}
\dot{V}_1(t) \leqslant & \sum_{i=1}^{m} \alpha_i \hat{e}_i^T(t) \Big\{ \left[\Psi_i \otimes (I_n + \Theta)\right] \hat{e}_i(t) + 2 \sum_{r=1}^{\tau} \sum_{j=i+1}^{m} c_r \big((\Psi_i \bar{G}^r_{ij}) \otimes \Gamma_r\big) \hat{e}_j(t) \\
& + 2 \sum_{r=1}^{\tau} c_r \big((\Psi_i \tilde{G}^r_i) \otimes \Gamma_r\big) \hat{e}_i(t) \Big\} \\
= & e^T(t) \Big[\hat{\Psi} \otimes (I_n + \Theta) + \sum_{r=1}^{\tau} c_r \big(\hat{\Psi} \hat{G}^r + (\hat{G}^r)^T \hat{\Psi}\big) \otimes \Gamma_r \Big] e(t) \\
\leqslant & \varphi_1 \|e(t)\|^2,
\end{aligned}
\tag{8.12}
$$

where $\hat{\Psi} = \mathrm{diag}(\alpha_1 \Psi_1, \alpha_2 \Psi_2, \cdots, \alpha_m \Psi_m)$ and $\varphi_1 = \lambda_M\big(\hat{\Psi} \otimes (I_n + \Theta) + \sum_{r=1}^{\tau} c_r (\hat{\Psi} \hat{G}^r + (\hat{G}^r)^T \hat{\Psi}) \otimes \Gamma_r\big)$.

From the definition of $V_1(t)$, we have

$$
\varphi_2 \|e(t)\|^2 \leqslant V_1(t) \leqslant \varphi_3 \|e(t)\|^2, \tag{8.13}
$$

where $\varphi_2 = \min_{i=1,2,\cdots,m}\{\lambda_m(\alpha_i \Psi_i)\}$ and $\varphi_3 = \max_{i=1,2,\cdots,m}\{\lambda_M(\alpha_i \Psi_i)\}$

Hence, by (8.12) and (8.13), one obtains

$$
\dot{V}_1(t) \leqslant \frac{\varphi_1}{\varphi_3} V_1(t). \tag{8.14}
$$

It follows from (8.13) and (8.14) that:

$$
\|e(t)\| \leqslant \sqrt{\frac{\varphi_3}{\varphi_2}} e^{\frac{\varphi_1}{2\varphi_3} t} \|e(0)\|. \tag{8.15}
$$

Consequently, we can conclude that the network (8.3) is synchronized. □

Let

$$\mathcal{D}^{r,i}=\begin{pmatrix}\mathcal{D}^{r,i}_{11} & \mathcal{D}^{r,i}_{12} & \cdots & \mathcal{D}^{r,i}_{1i}\\ (\mathcal{D}^{r,i}_{12})^T & \mathcal{D}^{r,i}_{22} & \cdots & \mathcal{D}^{r,i}_{2i}\\ \vdots & \vdots & \ddots & \vdots\\ (\mathcal{D}^{r,i}_{1i})^T & (\mathcal{D}^{r,i}_{2i})^T & \cdots & \mathcal{D}^{r,i}_{ii}\end{pmatrix},$$

where $\mathbb{R}\ni\alpha_i>0, i=1,2,\cdots,m$, and

$$\begin{aligned}
\mathcal{D}^{r,i}_{11}&=\alpha_1\Psi_1(c_r\tilde{G}^r_1+\xi I_{p_1})+\alpha_1(c_r\tilde{G}^r_1+\xi I_{p_1})^T\Psi_1,\\
\mathcal{D}^{r,i}_{12}&=\alpha_1 c_r\Psi_1\bar{G}^r_{12},\\
&\vdots\\
\mathcal{D}^{r,i}_{1i}&=\alpha_1 c_r\Psi_1\bar{G}^r_{1i},\\
\mathcal{D}^{r,i}_{22}&=\alpha_2\Psi_2(c_r\tilde{G}^r_2+\xi I_{p_2})+\alpha_2(c_r\tilde{G}^r_2+\xi I_{p_2})^T\Psi_2,\\
&\vdots\\
\mathcal{D}^{r,i}_{2i}&=\alpha_2 c_r\Psi_2\bar{G}^r_{2i},\\
&\vdots\\
\mathcal{D}^{r,i}_{ii}&=\alpha_i\Psi_i(c_r\tilde{G}^r_i+\xi I_{p_i})+\alpha_i(c_r\tilde{G}^r_i+\xi I_{p_i})^T\Psi_i.
\end{aligned}$$

The following corollary can be easily derived from Theorem 8.3.

Corollary 8.4 *If there exist matrices* $0<\Psi_i=\mathrm{diag}(\Psi_{i1},\Psi_{i2},\cdots,\Psi_{ip_i})\in\mathbb{R}^{p_i\times p_i}$, $i=1,2,\cdots,m$, *such that*

$$\Psi_i(c_r\tilde{G}^r_i+\xi I_{p_i})+(c_r\tilde{G}^r_i+\xi I_{p_i})^T\Psi_i<0, \tag{8.16}$$

where $r=1,2,\cdots,\tau$, *then the network* (8.3) *is synchronized.*

Proof By using the definition of $\mathcal{D}^{r,i}$, one has

$$\mathcal{D}^{r,m}=\boldsymbol{\alpha}\Psi(c_r\hat{G}^r+\xi I_N)+(c_r\hat{G}^r+\xi I_N)^T\Psi\boldsymbol{\alpha},$$

where $\boldsymbol{\alpha}=\mathrm{diag}(\alpha_1 I_{p_1},\alpha_2 I_{p_2},\cdots,\alpha_m I_{p_m})$, $\Psi=\mathrm{diag}(\Psi_1,\Psi_2,\cdots,\Psi_m)$.

From (8.16), it is easy to obtain $\mathcal{D}^{r,1}<0$. We can assume $\mathcal{D}^{r,i}<0$ for any $i\in\{1,2,\cdots,m-1\}$. Then, we shall demonstrate that $\mathcal{D}^{r,i+1}<0$. By utilizing Schur's Lemma, $\mathcal{D}^{r,i+1}<0$ is equivalent to

$$\mathcal{D}^{r,i}-\frac{1}{\alpha_{i+1}}\mathcal{F}\mathcal{Q}^{-1}\mathcal{F}^T<0, \tag{8.17}$$

where $\mathcal{Q} = \Psi_{i+1}(c_r\tilde{G}^r_{i+1} + \xi I_{p_{i+1}}) + (c_r\tilde{G}^r_{i+1} + \xi I_{p_{i+1}})^T\Psi_{i+1}$, $\mathcal{F}^T = (\alpha_1 c_r(\bar{G}^r_{1(i+1)})^T\Psi_1, \alpha_2 c_r(\bar{G}^r_{2(i+1)})^T\Psi_2, \cdots, \alpha_i c_r(\bar{G}^r_{i(i+1)})^T\Psi_i)$.

Obviously, by selecting α_{i+1} to be sufficiently larger than α_j for $j < i+1$, (8.17) can be satisfied. In view of the mathematical induction, we can easily derived that $\mathcal{D}^{r,m} < 0$. Thus, if (8.16) is satisfied, then there exists a matrix $0 < \boldsymbol{\alpha} = \text{diag}(\alpha_1 I_{p_1}, \alpha_2 I_{p_2}, \cdots, \alpha_m I_{p_m})$ such that

$$\boldsymbol{\alpha}\Psi(c_r\hat{G}^r + \xi I_N) + (c_r\hat{G}^r + \xi I_N)^T\Psi\boldsymbol{\alpha} < 0, r = 1, 2, \cdots, \tau.$$

On the other hand,

$$\begin{aligned}
&(\boldsymbol{\alpha}\Psi)\otimes(I_n+\Theta)+\sum_{r=1}^{\tau}c_r(\boldsymbol{\alpha}\Psi\hat{G}^r+(\hat{G}^r)^T\Psi\boldsymbol{\alpha})\otimes\Gamma_r\\
&\leqslant(\lambda_M(I_n+\Theta)\boldsymbol{\alpha}\Psi)\otimes I_n+\sum_{r=1}^{\tau}c_r(\boldsymbol{\alpha}\Psi\hat{G}^r+(\hat{G}^r)^T\Psi\boldsymbol{\alpha})\otimes\Gamma_r\\
&\leqslant(\lambda_M(I_n+\Theta)\boldsymbol{\alpha}\Psi)\otimes\frac{\sum_{r=1}^{\tau}\Gamma_r}{\sum_{r=1}^{\tau}\lambda_m(\Gamma_r)}+\sum_{r=1}^{\tau}c_r(\boldsymbol{\alpha}\Psi\hat{G}^r+(\hat{G}^r)^T\Psi\boldsymbol{\alpha})\otimes\Gamma_r\\
&=\sum_{r=1}^{\tau}(2\xi\boldsymbol{\alpha}\Psi)\otimes\Gamma_r+\sum_{r=1}^{\tau}c_r(\boldsymbol{\alpha}\Psi\hat{G}^r+(\hat{G}^r)^T\Psi\boldsymbol{\alpha})\otimes\Gamma_r\\
&=\sum_{r=1}^{\tau}\Big(\boldsymbol{\alpha}\Psi(c_r\hat{G}^r+\xi I_N)+(c_r\hat{G}^r+\xi I_N)^T\Psi\boldsymbol{\alpha}\Big)\otimes\Gamma_r\\
&<0.
\end{aligned}$$

Therefore, we can obtain from Theorem 8.3 that the network (8.3) is synchronized. □

From Corollary 8.4, it is easy to obtain the following conclusion.

Corollary 8.5 *If there exist matrices* $0 < \Psi_i = \text{diag}(\Psi_{i1}, \Psi_{i2}, \cdots, \Psi_{ip_i}) \in \mathbb{R}^{p_i \times p_i}$, $i = 1, 2, \cdots, m$, *such that*

$$c_r > \max_{i=1,2,\cdots,m}\left\{\frac{-2\xi\lambda_M(\Psi_i)}{\lambda_M(\Psi_i\tilde{G}^r_i+(\tilde{G}^r_i)^T\Psi_i)}\right\}, \quad (8.18)$$

where $r = 1, 2, \cdots, \tau$, $\Psi_i\tilde{G}^r_i + (\tilde{G}^r_i)^T\Psi_i < 0$, *then the network* (8.3) *achieves synchronization.*

Remark 8.6 From Corollary 8.5, there exist critical coupling strengths $c^*_r, r = 1, 2, \cdots, \tau$, for given $\Psi_i, i = 1, 2, \cdots, m$, such that the network (8.3) will synchronize if $c_r > c^*_r$ for all $r = 1, 2, \cdots, \tau$. Thus, if $\Psi_i\tilde{G}^r_i + (\tilde{G}^r_i)^T\Psi_i < 0, i =$

$1, 2, \cdots, m$, are satisfied, then the network (8.3) can achieve synchronization as long as the coupling strengths $c_r, r = 1, 2, \cdots, \tau$, are large enough.

8.2.1.3 Pinning Adaptive Synchronization Criteria

For (8.18), the coupling strengths $c_r, r = 1, 2, \cdots, \tau$, are very conservative and usually much larger than the needed values [29, 37, 38]. Consequently, we shall design some adaptive strategies to tune the coupling strengths $c_r, r = 1, 2, \cdots, \tau$, of network (8.3) in this subsection.

Theorem 8.7 *If there exist matrices* $0 < \Psi_i = \text{diag}(\Psi_{i1}, \Psi_{i2}, \cdots, \Psi_{ip_i}) \in \mathbb{R}^{p_i \times p_i}$, $i = 1, 2, \cdots, m$, *such that*

$$\Psi_i \tilde{G}_i^r + (\tilde{G}_i^r)^T \Psi_i < 0, \tag{8.19}$$

where $r = 1, 2, \cdots, \tau$, *then the network* (8.3) *achieves synchronization under the following adaptive laws:*

$$\dot{c}_r(t) = \beta_r \sum_{i=1}^{m} \alpha_i \hat{e}_i^T(t)(\Psi_i \otimes \Gamma_r)\hat{e}_i(t), \tag{8.20}$$

where $r = 1, 2, \cdots, \tau, 0 < \beta_r \in \mathbb{R}$, *and* $\alpha_i, i = 1, 2, \cdots, m$, *are positive real numbers satisfying* (8.23).

Proof The network (8.8) with adaptive coupling strengths is characterized by

$$\begin{aligned}\dot{\hat{e}}_i(t) =& \hat{f}_i(x(t)) - \hat{f}_i(s^*(t)) + \sum_{r=1}^{\tau}\left[c_r(t) \sum_{j=i+1}^{m} (\bar{G}_{ij}^r \otimes \Gamma_r)\hat{e}_j(t)\right.\\ &\left. + c_r(t)(\tilde{G}_i^r \otimes \Gamma_r)\hat{e}_i(t)\right], \quad i = 1, 2, \cdots, m.\end{aligned} \tag{8.21}$$

From (8.19), there exist $\mathbb{R} \ni b_r > 0, r = 1, 2, \cdots, \tau$ such that

$$\Psi_i(\tilde{G}_i^r + b_r I_{p_i}) + (\tilde{G}_i^r + b_r I_{p_i})^T \Psi_i < 0, \tag{8.22}$$

where $i = 1, 2, \cdots, m$.

Choose the following Lyapunov functional for the system (8.21):

$$V_2(t) = \sum_{i=1}^{m} \alpha_i \hat{e}_i^T(t)(\Psi_i \otimes I_n)\hat{e}_i(t) + \sum_{r=1}^{\tau} \frac{b_r}{\beta_r}(c_r(t) - \hat{c}_r)^2,$$

where $\hat{c}_r > 0, r = 1, 2, \cdots, \tau$.

Then, one has

$$\dot{V}_2(t) \leqslant e^T(t)\Big[(\boldsymbol{\alpha}\Psi)\otimes(I_n+\Theta)+\sum_{r=1}^{\tau}c_r(t)\big(\boldsymbol{\alpha}\Psi\hat{G}^r+(\hat{G}^r)^T\Psi\boldsymbol{\alpha}+2b_r\boldsymbol{\alpha}\Psi\big)\otimes\Gamma_r$$
$$-\sum_{r=1}^{\tau}2b_r\hat{c}_r\left((\boldsymbol{\alpha}\Psi)\otimes\Gamma_r\right)\Big]e(t),$$

where $\boldsymbol{\alpha} = \mathrm{diag}(\alpha_1 I_{p_1}, \alpha_2 I_{p_2}, \cdots, \alpha_m I_{p_m})$, $\Psi = \mathrm{diag}(\Psi_1, \Psi_2, \cdots, \Psi_m)$, $e(t) = (\hat{e}_1^T(t), \hat{e}_2^T(t), \cdots, \hat{e}_m^T(t))^T$.

Then, following similar arguments as in the proof of Corollary 8.4, we can find a matrix $0 < \hat{\boldsymbol{\alpha}} = \mathrm{diag}(\hat{\alpha}_1 I_{p_1}, \hat{\alpha}_2 I_{p_2}, \cdots, \hat{\alpha}_m I_{p_m})$ satisfying

$$\hat{\boldsymbol{\alpha}}\Psi\hat{G}^r+(\hat{G}^r)^T\Psi\hat{\boldsymbol{\alpha}}+2b_r\hat{\boldsymbol{\alpha}}\Psi < 0, \tag{8.23}$$

where $r = 1, 2, \cdots, \tau$.

Let $\boldsymbol{\alpha} = \hat{\boldsymbol{\alpha}}$. Because $c_r(t) > 0$ and $\Gamma_r > 0$, one has

$$\dot{V}_2(t) \leqslant e^T(t)\Big[(\boldsymbol{\alpha}\Psi)\otimes(I_n+\Theta)-\sum_{r=1}^{\tau}2b_r\hat{c}_r\left((\boldsymbol{\alpha}\Psi)\otimes\Gamma_r\right)\Big]e(t).$$

By choosing $\hat{c}_r$ sufficiently large such that

$$I_n+\Theta-\sum_{r=1}^{\tau}2b_r\hat{c}_r\Gamma_r < 0,$$

one gets

$$\dot{V}_2(t) \leqslant -\varrho\|e(t)\|^2, \tag{8.24}$$

where $\varrho = \lambda_m\big((\boldsymbol{\alpha}\Psi)\otimes(-I_n-\Theta+\sum_{r=1}^{\tau}2b_r\hat{c}_r\Gamma_r)\big) > 0$.

It is obvious that $V_2(t)$ is non-increasing, and each term in $V_2(t)$ is bounded. Thus, we know that $c_r(t), r = 1, 2, \cdots, \tau$, are bounded, and $\lim_{t\to+\infty} V_2(t)$ exists and is a non-negative real number. On the other hand, $c_r(t), r = 1, 2, \cdots, \tau$, are monotonically increasing (see (8.20)), thus we can derive that $c_r(t), r = 1, 2, \cdots, \tau$, asymptotically converge to some finite positive values. Hence, based on the definition of $V_2(t)$, we can obtain that $\lim_{t\to+\infty}\sum_{i=1}^{m}\alpha_i\hat{e}_i^T(t)(\Psi_i\otimes I_n)\hat{e}_i(t)$ exists and is a non-negative real number.

Based on a contradiction method, we shall prove

$$\lim_{t\to+\infty}\sum_{i=1}^{m}\alpha_i\hat{e}_i^T(t)(\Psi_i\otimes I_n)\hat{e}_i(t) = 0.$$

If this is not true, one obtains

$$\lim_{t\to+\infty}\sum_{i=1}^{m}\alpha_i\hat{e}_i^T(t)(\Psi_i\otimes I_n)\hat{e}_i(t)=\sigma>0.$$

Then, there exists a real number $\Delta>0$ such that $\sum_{i=1}^{m}\alpha_i\hat{e}_i^T(t)(\Psi_i\otimes I_n)\hat{e}_i(t)>\frac{\sigma}{2}$ for $t\geqslant\Delta$. Consequently,

$$\|e(t)\|^2>\frac{\sigma}{2\varphi_3},\quad t\geqslant\Delta, \tag{8.25}$$

where $\varphi_3=\max_{i=1,2,\cdots,m}\{\lambda_M(\alpha_i\Psi_i)\}$. According to (8.24) and (8.25), one has

$$\dot{V}_2(t)<-\frac{\varrho\sigma}{2\varphi_3},\quad t\geqslant\Delta. \tag{8.26}$$

By (8.26), we can get

$$\begin{aligned}-V_2(\Delta)&\leqslant V_2(+\infty)-V_2(\Delta)\\&=\int_{\Delta}^{+\infty}\dot{V}_2(t)dt\\&<-\int_{\Delta}^{+\infty}\frac{\varrho\sigma}{2\varphi_3}dt\\&=-\infty,\end{aligned}$$

which is a contradiction. Thus, one has

$$\lim_{t\to+\infty}\sum_{i=1}^{m}\alpha_i\hat{e}_i^T(t)(\Psi_i\otimes I_n)\hat{e}_i(t)=0.$$

Then, it is easy to obtain $\lim_{t\to+\infty}\|e(t)\|=0$. Hence, the network (8.3) achieves synchronization under the adaptive laws (8.20). □

In what follows, we further discuss the pinning synchronization of multi-weighted complex network (8.3) with adaptive feedback gains. Design the following adaptive laws:

$$\dot{h}_i^r(t)=\hat{\beta}_r(x_{q_i}(t)-\hat{s}(t))^T\Gamma_r(x_{q_i}(t)-\hat{s}(t)), \tag{8.27}$$

where $\hat{\beta}_r$ and $h_i^r(0)$ are positive real numbers, $i=1,2,\cdots,l$, $r=1,2,\cdots,\tau$.

By applying (8.3), (8.4) and (8.27), we have

$$\dot{\hat{e}}_i(t) = \hat{f}_i(x(t)) - \hat{f}_i(s^*(t)) + \sum_{r=1}^{\tau}\sum_{j=i+1}^{m} c_r(\bar{G}_{ij}^r \otimes \Gamma_r)\hat{e}_j(t)$$
$$- \sum_{r=1}^{\tau} c_r(H_i^r(t) \otimes \Gamma_r)\hat{e}_i(t) + \sum_{r=1}^{\tau} c_r(\bar{G}_i^r \otimes \Gamma_r)\hat{e}_i(t), \qquad (8.28)$$

where $i = 1, 2, \cdots, m, \hat{e}_i(t), \hat{f}_i(x(t)), \hat{f}_i(s^*(t)), \bar{G}_{ij}^r, \bar{G}_i^r$ represent the same meanings as those in (8.7) and (8.8), $H_i^r(t) = \mathrm{diag}(H_{i1}^r(t), H_{i2}^r(t), \cdots, H_{ip_i}^r(t)) \in \mathbb{R}^{p_i \times p_i}$, $H^r(t) = \mathrm{diag}(H_1^r(t), H_2^r(t), \cdots, H_m^r(t)) = \mathrm{diag}(0, \cdots, 0, \underbrace{h_1^r(t), 0, \cdots, 0,}_{q_1}$ $\underbrace{h_2^r(t), 0, \cdots, 0,}_{q_2} \underbrace{h_l^r(t), \cdots}_{q_l})$.

Theorem 8.8 *If there exist matrices* $\Psi_i = \mathrm{diag}(\Psi_{i1}, \Psi_{i2}, \cdots, \Psi_{ip_i}) \in \mathbb{R}^{p_i \times p_i} > 0, \hat{\kappa}_i^r = \mathrm{diag}(\kappa_{i1}^r, \kappa_{i2}^r, \cdots, \kappa_{ip_i}^r) \in \mathbb{R}^{p_i \times p_i} \geqslant 0$ *and positive real numbers* $\alpha_i, r = 1, 2, \cdots, \tau, i = 1, 2, \cdots, m$, *such that*

$$(\boldsymbol{\alpha}\Psi) \otimes W + \sum_{r=1}^{\tau} c_r\left(\boldsymbol{\alpha}\Psi G^r + (G^r)^T \Psi \boldsymbol{\alpha} - 2\boldsymbol{\alpha}\Psi\hat{\kappa}^r\right) \otimes \Gamma_r < 0 \qquad (8.29)$$

where $\boldsymbol{\alpha} = \mathrm{diag}(\alpha_1 I_{p_1}, \alpha_2 I_{p_2}, \cdots, \alpha_m I_{p_m}), \Psi = \mathrm{diag}(\Psi_1, \Psi_2, \cdots, \Psi_m), W = I_n + \Theta, \hat{\kappa}^r = \mathrm{diag}(\hat{\kappa}_1^r, \hat{\kappa}_2^r, \cdots, \hat{\kappa}_m^r), \kappa_{ij}^r = 0 (j = 1, 2, \cdots, p_i)$ *if and only if* $H_{ij}^r(t) = 0$, *then the network* (8.3) *can achieve synchronization under the adaptive laws* (8.27).

Proof Firstly, we select the following Lyapunov functional for system (8.28):

$$V_3(t) = \sum_{i=1}^{m} \alpha_i \hat{e}_i^T(t)(\Psi_i \otimes I_n)\hat{e}_i(t)$$
$$+ \sum_{r=1}^{\tau}\sum_{i=1}^{m} \frac{c_r\alpha_i}{\hat{\beta}_r}(\hat{H}_i^r(t) - \kappa_i^r)^T \Psi_i(\hat{H}_i^r(t) - \kappa_i^r),$$

where $\hat{H}_i^r(t) = (H_{i1}^r(t), H_{i2}^r(t), \cdots, H_{ip_i}^r(t))^T, \kappa_i^r = (\kappa_{i1}^r, \kappa_{i2}^r, \cdots, \kappa_{ip_i}^r)^T$.

Then, one obtains

$$\dot{V}_3(t) = 2\sum_{i=1}^{m} \alpha_i \hat{e}_i^T(t)(\Psi_i \otimes I_n)\Big[\hat{f}_i(x(t)) - \hat{f}_i(s^*(t))$$
$$+ \sum_{r=1}^{\tau}\sum_{j=i+1}^{m} c_r(\bar{G}_{ij}^r \otimes \Gamma_r)\hat{e}_j(t) + \sum_{r=1}^{\tau} c_r(\bar{G}_i^r \otimes \Gamma_r)\hat{e}_i(t)$$

$$
\begin{aligned}
&-\sum_{r=1}^{\tau} c_r (H_i^r(t) \otimes \Gamma_r)\hat{e}_i(t)\Big]\\
&+2\sum_{r=1}^{\tau}\sum_{i=1}^{m} c_r\alpha_i \hat{e}_i^T(t)\Big[(\Psi_i H_i^r(t) - \Psi_i \hat{\kappa}_i^r)\otimes \Gamma_r\Big]\hat{e}_i(t)\\
&\leqslant e^T(t)\Big[(\alpha\Psi)\otimes(I_n+\Theta)+\sum_{r=1}^{\tau} c_r\big(\alpha\Psi G^r\\
&+(G^r)^T\Psi\alpha - 2\alpha\Psi\hat{\kappa}^r\big)\otimes\Gamma_r\Big]e(t),
\end{aligned}
$$

where $e(t) = (e_1^T(t), e_2^T(t), \cdots, e_N^T(t))^T$.

By utilizing similar arguments as in Theorem 8.7, one has

$$\lim_{t\to+\infty} e_i(t) = 0, i = 1, 2, \cdots, N.$$

Thus, the network (8.3) is synchronized under the adaptive laws (8.27). □

8.2.2 *Pinning Synchronization for Undirected Complex Dynamical Network with Multi-weights*

8.2.2.1 Network Model

When the topological structure of complex network (8.1) is undirected, we have

$$
\begin{aligned}
\dot{x}_i(t) = {} & f(x_i(t)) + c_1\sum_{j=1}^{N} G_{ij}^1\Gamma_1 x_j(t) + c_2\sum_{j=1}^{N} G_{ij}^2\Gamma_2 x_j(t) + \cdots\\
&+c_\tau\sum_{j=1}^{N} G_{ij}^\tau\Gamma_\tau x_j(t),
\end{aligned}
\tag{8.30}
$$

where $i = 1, 2, \cdots, N$, $G^r = (G_{ij}^r)_{N\times N}$ $(r = 1, 2, \cdots, \tau)$ is the coupling configuration matrix representing coupling weights in the rth coupling form. We can define G_{ij}^r as follows: if there exists a connection between nodes i and j, then $G_{ij}^r = G_{ji}^r > 0$; otherwise, $G_{ij}^r = G_{ji}^r = 0 (i \neq j)$. The diagonal elements of matrix G^r satisfy

$$G_{ii}^r = -\sum_{\substack{j=1\\ j\neq i}}^{N} G_{ij}^r, i = 1, 2, \cdots, N.$$

In this section, they also have the same topology structure for different coupling forms.

Suppose that G^r is in the Frobenius normal form, that is,

$$G^r = \begin{pmatrix} \bar{G}_1^r & 0 & \cdots & 0 \\ 0 & \bar{G}_2^r & \cdots & 0 \\ \vdots & \vdots & \ddots & \vdots \\ 0 & 0 & \cdots & \bar{G}_m^r \end{pmatrix}, \quad r=1,2,\cdots,\tau, \tag{8.31}$$

where $\bar{G}_1^r \in \mathbb{R}^{p_1 \times p_1}, \bar{G}_2^r \in \mathbb{R}^{p_2 \times p_2}, \cdots, \bar{G}_m^r \in \mathbb{R}^{p_m \times p_m}$ are square irreducible matrices.

We can apply the pinning strategy to a small fraction of the nodes in network (8.30) for achieving synchronization. Assume that the nodes $q_1, q_2, \cdots, q_l (1 \leqslant l < N)$ are chosen to be pinned. After that, the pinning controlled network is given by

$$\begin{aligned} \dot{x}_{q_i}(t) &= f(x_{q_i}(t)) + c_1 \sum_{j=1}^{N} G_{q_i j}^1 \Gamma_1 x_j(t) + c_2 \sum_{j=1}^{N} G_{q_i j}^2 \Gamma_2 x_j(t) + \cdots \\ &\quad + c_\tau \sum_{j=1}^{N} G_{q_i j}^\tau \Gamma_\tau x_j(t) + u_{q_i}(t), \quad i = 1, 2, \cdots, l, \\ \dot{x}_{q_i}(t) &= f(x_{q_i}(t)) + c_1 \sum_{j=1}^{N} G_{q_i j}^1 \Gamma_1 x_j(t) + c_2 \sum_{j=1}^{N} G_{q_i j}^2 \Gamma_2 x_j(t) + \cdots \\ &\quad + c_\tau \sum_{j=1}^{N} G_{q_i j}^\tau \Gamma_\tau x_j(t), \quad i = l+1, l+2, \cdots, N, \end{aligned} \tag{8.32}$$

where

$$u_{q_i}(t) = -\sum_{r=1}^{\tau} c_r h_i^r \Gamma_r (x_{q_i}(t) - \hat{s}(t)), \quad i=1,2,\cdots,l, \tag{8.33}$$

are n-dimensional linear feedback controllers with all the control gains $\mathbb{R} \ni h_i^r > 0$, $\hat{s}(t)$ denotes the same meaning as it in subsection 8.2.1.

By applying controllers (8.33) into system (8.32), one gets

$$\dot{x}_i(t) = f(x_i(t)) + \sum_{r=1}^{\tau} \sum_{j=1}^{N} c_r G_{ij}^r \Gamma_r x_j(t) - \sum_{r=1}^{\tau} c_r \hat{h}_i^r \Gamma_r (x_i(t) - \hat{s}(t)), \tag{8.34}$$

where $i = 1, 2, \cdots, N$,

$$H^r = \text{diag}(\hat{h}_1^r, \hat{h}_2^r, \cdots, \hat{h}_N^r)$$
$$= \text{diag}(0, \cdots, 0, \underbrace{h_1^r}_{q_1}, 0, \cdots, 0, \underbrace{h_2^r}_{q_2}, 0, \cdots, 0, \underbrace{h_l^r}_{q_l}, \cdots), r = 1, 2, \cdots, \tau.$$

Let $e_i(t) = (e_{i1}(t), e_{i2}(t), \cdots, e_{in}(t))^T = x_i(t) - \hat{s}(t)$. Then, one obtains

$$\dot{e}_i(t) = f(x_i(t)) - f(\hat{s}(t)) + \sum_{r=1}^{\tau}\sum_{j=1}^{N} c_r G_{ij}^r \Gamma_r e_j(t) - \sum_{r=1}^{\tau} c_r \hat{h}_i^r \Gamma_r e_i(t),$$
$$i = 1, 2, \cdots, N. \tag{8.35}$$

8.2.2.2 Pinning Synchronization Criteria

Define $H^r = \text{diag}(H_1^r, H_2^r, \cdots, H_m^r)$, $H_i^r \in \mathbb{R}^{p_i \times p_i}$, $i = 1, 2, \cdots, m$, and $\hat{G}^r = G^r - H^r$. Then, one has

$$\hat{G}^r = \begin{pmatrix} \tilde{G}_1^r & 0 & \cdots & 0 \\ 0 & \tilde{G}_2^r & \cdots & 0 \\ \vdots & \vdots & \ddots & \vdots \\ 0 & 0 & \cdots & \tilde{G}_m^r \end{pmatrix}, \quad r = 1, 2, \cdots, \tau, \tag{8.36}$$

where $\tilde{G}_1^r \in \mathbb{R}^{p_1 \times p_1} = \bar{G}_1^r - H_1^r$, $\tilde{G}_2^r \in \mathbb{R}^{p_2 \times p_2} = \bar{G}_2^r - H_2^r, \cdots, \tilde{G}_m^r \in \mathbb{R}^{p_m \times p_m} = \bar{G}_m^r - H_m^r$ are square irreducible matrices.

Then, the system (8.35) can be rewritten as

$$\dot{\hat{e}}_i(t) = \hat{f}_i(x(t)) - \hat{f}_i(s^*(t)) + \sum_{r=1}^{\tau} c_r (\tilde{G}_i^r \otimes \Gamma_r) \hat{e}_i(t), \tag{8.37}$$

where $i = 1, 2, \cdots, m$, $\hat{e}_i(t)$, $s^*(t)$, $\hat{f}_i(s^*(t))$, $\hat{f}_i(x(t))$ represent the same meanings as those in (8.8).

Theorem 8.9 *If there exist matrices* $0 < \Psi_i = \text{diag}(\Psi_{i1}, \Psi_{i2}, \cdots, \Psi_{ip_i}) \in \mathbb{R}^{p_i \times p_i}$, $i = 1, 2, \cdots, m$, *such that*

$$\Psi_i (c_r \tilde{G}_i^r + \xi I_{p_i}) + (c_r \tilde{G}_i^r + \xi I_{p_i}) \Psi_i < 0, \tag{8.38}$$

where $r = 1, 2, \cdots, \tau$, $\Theta = \text{diag}(\rho_1^2, \rho_2^2, \cdots, \rho_n^2)$, $\xi = \frac{\lambda_M(I_n + \Theta)}{2\sum_{r=1}^{\tau}\lambda_m(\Gamma_r)}$, $W = I_n + \Theta$, *then the network* (8.32) *is synchronized.*

Proof We choose the following Lyapunov functional for system (8.37):

$$\tilde{V}_i(t) = \hat{e}_i^T(t)(\Psi_i \otimes I_n)\hat{e}_i(t), \ i = 1, 2, \cdots, m. \tag{8.39}$$

Then, one has,

$$\begin{aligned}
\dot{\tilde{V}}_i(t) &\leqslant \hat{e}_i^T(t)\left[\Psi_i \otimes (I_n + \Theta) + \sum_{r=1}^{\tau} c_r\big(\Psi_i \tilde{G}_i^r + \tilde{G}_i^r \Psi_i\big) \otimes \Gamma_r\right]\hat{e}_i(t) \\
&\leqslant \hat{e}_i^T(t)\left[(2\xi\Psi_i) \otimes \left(\sum_{r=1}^{\tau}\Gamma_r\right) + \sum_{r=1}^{\tau} c_r\big(\Psi_i \tilde{G}_i^r + \tilde{G}_i^r \Psi_i\big) \otimes \Gamma_r\right]\hat{e}_i(t) \\
&= \hat{e}_i^T(t)\left\{\sum_{r=1}^{\tau}\left[\Psi_i(c_r\tilde{G}_i^r + \xi I_{p_i}) + (c_r\tilde{G}_i^r + \xi I_{p_i})\Psi_i\right] \otimes \Gamma_r\right\}\hat{e}_i(t) \\
&\leqslant \hat{\varphi}_i \|\hat{e}_i(t)\|^2,
\end{aligned} \tag{8.40}$$

where $\hat{\varphi}_i = \lambda_M\left(\sum_{r=1}^{\tau}\left[\Psi_i(c_r\tilde{G}_i^r + \xi I_{p_i}) + (c_r\tilde{G}_i^r + \xi I_{p_i})\Psi_i\right] \otimes \Gamma_r\right)$.

By using the definition of $\tilde{V}_i(t)$, we can derive

$$\hat{\eta}_i \|\hat{e}_i(t)\|^2 \leqslant \tilde{V}_i(t) \leqslant \bar{\eta}_i \|\hat{e}_i(t)\|^2, \tag{8.41}$$

where $\hat{\eta}_i = \lambda_m(\Psi_i)$ and $\bar{\eta}_i = \lambda_M(\Psi_i)$.

Hence, in view of (8.40) and (8.41), one has

$$\dot{\tilde{V}}_i(t) \leqslant \frac{\hat{\varphi}_i}{\bar{\eta}_i}\tilde{V}_i(t). \tag{8.42}$$

Then, we obtain from (8.41) and (8.42) that

$$\|\hat{e}_i(t)\| \leqslant \sqrt{\frac{\bar{\eta}_i}{\hat{\eta}_i}} e^{\frac{\hat{\varphi}_i}{2\bar{\eta}_i}t}\|\hat{e}_i(0)\|, \ \ i = 1, 2, \cdots, m. \tag{8.43}$$

Thus, the network (8.32) achieves synchronization. □

By Theorem 8.9, we can easily obtain the following corollary.

Corollary 8.10 *If there exist matrices* $0 < \Psi_i = \mathrm{diag}(\Psi_{i1}, \Psi_{i2}, \cdots, \Psi_{ip_i}) \in \mathbb{R}^{p_i \times p_i}$, $i = 1, 2, \cdots, m$, *such that*

$$c_r > \max_{i=1,2,\cdots,m}\left\{\frac{-2\xi\lambda_M(\Psi_i)}{\lambda_M(\Psi_i\tilde{G}_i^r + \tilde{G}_i^r\Psi_i)}\right\}, \tag{8.44}$$

where $\Psi_i\tilde{G}_i^r + \tilde{G}_i^r\Psi_i < 0, r = 1, 2, \cdots, \tau$, *then the network* (8.32) *is synchronized.*

8.2.2.3 Pinning Adaptive Synchronization Criteria

Theorem 8.11 *If there exist matrices* $0 < \Psi_i = \mathrm{diag}(\Psi_{i1}, \Psi_{i2}, \cdots, \Psi_{ip_i}) \in \mathbb{R}^{p_i \times p_i}$, $i = 1, 2, \cdots, m$, *such that*

$$\Psi_i \tilde{G}_i^r + \tilde{G}_i^r \Psi_i < 0, \tag{8.45}$$

then the network (8.32) *can achieve synchronization under the following adaptive laws:*

$$\dot{c}_r(t) = \sum_{i=1}^{m} \varepsilon_r \hat{e}_i^T(t)(\Psi_i \otimes \Gamma_r)\hat{e}_i(t), \tag{8.46}$$

where $0 < \varepsilon_r \in \mathbb{R}, r = 1, 2, \cdots, \tau$.

Proof The network (8.37) with adaptive coupling strengths is characterized by

$$\dot{\hat{e}}_i(t) = \hat{f}_i(x(t)) - \hat{f}_i(s^*(t)) + \sum_{r=1}^{\tau} c_r(t)(\tilde{G}_i^r \otimes \Gamma_r)\hat{e}_i(t), \tag{8.47}$$

where $i = 1, 2, \cdots, m$.

From (8.45), there exist some positive constants $\tilde{b}_r, r = 1, 2, \cdots, \tau$, such that

$$\Psi_i(\tilde{G}_i^r + \tilde{b}_r I_{p_i}) + (\tilde{G}_i^r + \tilde{b}_r I_{p_i})\Psi_i < 0, \tag{8.48}$$

where $i = 1, 2, \cdots, m$.

For the system (8.47), we define the Lyapunov functional as

$$\bar{V}(t) = \sum_{i=1}^{m} \hat{e}_i^T(t)(\Psi_i \otimes I_n)\hat{e}_i(t) + \sum_{r=1}^{\tau} \frac{\tilde{b}_r}{\varepsilon_r}(c_r(t) - \hat{c}_r)^2,$$

where $\mathbb{R} \ni \hat{c}_r > 0, r = 1, 2, \cdots, \tau$.

Then, one gets

$$\begin{aligned}
\dot{\bar{V}}(t) \leqslant & \sum_{i=1}^{m} \hat{e}_i^T(t)\Big[\Psi_i \otimes (I_n + \Theta) + \sum_{r=1}^{\tau} c_r(t)\big(\Psi_i \tilde{G}_i^r \\
& + \tilde{G}_i^r \Psi_i + 2\tilde{b}_r \Psi_i\big) \otimes \Gamma_r - \sum_{r=1}^{\tau} 2\tilde{b}_r \hat{c}_r (\Psi_i \otimes \Gamma_r)\Big]\hat{e}_i(t) \\
\leqslant & \sum_{i=1}^{m} \hat{e}_i^T(t)\Big[\Psi_i \otimes (I_n + \Theta) - \sum_{r=1}^{\tau} 2\tilde{b}_r \hat{c}_r (\Psi_i \otimes \Gamma_r)\Big]\hat{e}_i(t).
\end{aligned}$$

By choosing $\hat{c}_r, r = 1, 2, \cdots, \tau$, sufficiently large such that

$$I_n + \Theta - \sum_{r=1}^{\tau} 2\tilde{b}_r \hat{c}_r \Gamma_r < 0,$$

one has

$$\dot{\hat{V}}(t) \leqslant -\hat{\varrho} \sum_{i=1}^{m} \|\hat{e}_i(t)\|^2,$$

where $\hat{\varrho} = \min_{i=1,2,\cdots,m} \left\{\lambda_m(\Psi_i \otimes (-I_n - \Theta + \sum_{r=1}^{\tau} 2\tilde{b}_r \hat{c}_r \Gamma_r))\right\} > 0$.

Then, by using similar arguments as in Theorem 8.7, one obtains

$$\lim_{t \to +\infty} e_i(t) = 0, i = 1, 2, \cdots, N.$$

Therefore, the network (8.32) achieves synchronization under the adaptive laws (8.46). □

In what follows, we further consider the synchronization of multi-weighted complex network (8.32) with adaptive feedback gains. Design the following adaptive laws:

$$\dot{h}_i^r(t) = \hat{\varepsilon}_r (x_{q_i}(t) - \hat{s}(t))^T \Gamma_r (x_{q_i}(t) - \hat{s}(t)), \tag{8.49}$$

where $\hat{\varepsilon}_r$ and $h_i^r(0)$ are positive constants, $i = 1, 2, \cdots, l, r = 1, 2, \cdots, \tau$.

By applying (8.32) and (8.49), one has

$$\begin{aligned}\dot{\hat{e}}_i(t) = {} & \hat{f}_i(x(t)) - \hat{f}_i(s^*(t)) - \sum_{r=1}^{\tau} c_r (H_i^r(t) \otimes \Gamma_r)\hat{e}_i(t) \\ & + \sum_{r=1}^{\tau} c_r (\bar{G}_i^r \otimes \Gamma_r)\hat{e}_i(t),\end{aligned} \tag{8.50}$$

where $i = 1, 2, \cdots, m$, $\hat{e}_i(t)$, $\hat{f}_i(x(t))$, $\hat{f}_i(s^*(t))$, $\bar{G}_i^r$ represent the same meanings as those in (8.36) and (8.37), $H_i^r(t) = \text{diag}(H_{i1}^r(t), H_{i2}^r(t), \cdots, H_{ip_i}^r(t)) \in \mathbb{R}^{p_i \times p_i}$, $H^r(t) = \text{diag}(H_1^r(t), H_2^r(t), \cdots, H_m^r(t)) = \text{diag}(0, \cdots, 0, \underbrace{h_1^r(t)}_{q_1}, 0, \cdots, 0, \underbrace{h_2^r(t)}_{q_2}, 0, \cdots, 0, \underbrace{h_l^r(t)}_{q_l}, \cdots)$.

Theorem 8.12 *If there exist matrices* $\Psi_i = \text{diag}(\Psi_{i1}, \Psi_{i2}, \cdots, \Psi_{ip_i}) \in \mathbb{R}^{p_i \times p_i} > 0$, $\hat{\kappa}_i^r = \text{diag}(\kappa_{i1}^r, \kappa_{i2}^r, \cdots, \kappa_{ip_i}^r) \in \mathbb{R}^{p_i \times p_i} \geqslant 0, r = 1, 2, \cdots, \tau, i = 1, 2, \cdots, m$, *such that*

$$\Psi \otimes W + \sum_{r=1}^{\tau} c_r \big(\Psi G^r + G^r \Psi - 2\Psi \hat{\kappa}^r\big) \otimes \Gamma_r < 0, \tag{8.51}$$

where $\Psi = \operatorname{diag}(\Psi_1, \Psi_2, \cdots, \Psi_m)$, $W = I_n + \Theta$, $\hat{\kappa}^r = \operatorname{diag}(\hat{\kappa}_1^r, \hat{\kappa}_2^r, \cdots, \hat{\kappa}_m^r)$, $\kappa_{ij}^r = 0 (j = 1, 2, \cdots, p_i)$ *if and only if* $H_{ij}^r(t) = 0$, *then the network* (8.32) *achieves synchronization under the adaptive laws* (8.49).

Proof Firstly, define the following Lyapunov functional for the system (8.50):

$$\begin{aligned}\bar{V}_1(t) &= \sum_{i=1}^{m} \hat{e}_i^T(t)(\Psi_i \otimes I_n)\hat{e}_i(t) \\ &\quad + \sum_{r=1}^{\tau}\sum_{i=1}^{m} \frac{c_r}{\hat{\varepsilon}_r}(\hat{H}_i^r(t) - \kappa_i^r)^T \Psi_i (\hat{H}_i^r(t) - \kappa_i^r),\end{aligned}$$

where $\hat{H}_i^r(t) = (H_{i1}^r(t), H_{i2}^r(t), \cdots, H_{ip_i}^r(t))^T, \kappa_i^r = (\kappa_{i1}^r, \kappa_{i2}^r, \cdots, \kappa_{ip_i}^r)^T$.

Then, one obtains

$$\begin{aligned}\dot{\bar{V}}_1(t) \leqslant e^T(t)\Big[&\Psi \otimes (I_n + \Theta) + \sum_{r=1}^{\tau} c_r \big(\Psi G^r \\ &+ G^r \Psi - 2\Psi \hat{\kappa}^r\big) \otimes \Gamma_r\Big] e(t),\end{aligned} \tag{8.52}$$

where $e(t) = (e_1^T(t), e_2^T(t), \cdots, e_N^T(t))^T$.

By carrying out similar arguments as in Theorem 8.8, one can easily derive

$$\lim_{t \to +\infty} e_i(t) = 0, i = 1, 2, \cdots, N.$$

Consequently, the network (8.32) is synchronized under the adaptive laws (8.49).
□

8.3 Numerical Examples

Example 8.13 Consider the following complex network with multi-weights and directed topology:

$$\begin{aligned}\dot{x}_i(t) = f(x_i(t)) &+ c_1 \sum_{j=1}^{10} G_{ij}^1 \Gamma_1 x_j(t) + c_2 \sum_{j=1}^{10} G_{ij}^2 \Gamma_2 x_j(t) \\ &+ c_3 \sum_{j=1}^{10} G_{ij}^3 \Gamma_3 x_j(t), \ \ i = 1, 2, \cdots, 10,\end{aligned} \tag{8.53}$$

where $f_l(\xi) = \frac{|\xi+1|-|\xi-1|}{4}, l = 1, 2, 3, \Gamma_1 = \mathrm{diag}(0.7, 0.6, 0.5), \Gamma_2 = \mathrm{diag}(0.8, 0.5, 0.6), \Gamma_3 = \mathrm{diag}(0.6, 0.7, 0.5), c_1 = 3, c_2 = 2, c_3 = 4$, the matrices $G^1 = (G^1_{ij})_{10\times10}$, $G^2 = (G^2_{ij})_{10\times10}$ and $G^3 = (G^3_{ij})_{10\times10}$ are chosen as, respectively

$$G^1 = \begin{pmatrix} -1.1 & 0.2 & 0.1 & 0.3 & 0.1 & 0.4 & 0 & 0 & 0 & 0 \\ 0.1 & -0.7 & 0.1 & 0.2 & 0.3 & 0 & 0 & 0 & 0 & 0 \\ 0.2 & 0.3 & -0.8 & 0.1 & 0.2 & 0 & 0 & 0 & 0 & 0 \\ 0.1 & 0.2 & 0.3 & -0.9 & 0.3 & 0 & 0 & 0 & 0 & 0 \\ 0.3 & 0.2 & 0.2 & 0.4 & -1.1 & 0 & 0 & 0 & 0 & 0 \\ 0 & 0 & 0 & 0 & 0 & -0.8 & 0.3 & 0.2 & 0.2 & 0.1 \\ 0 & 0 & 0 & 0 & 0 & 0.1 & -0.6 & 0.2 & 0.1 & 0.2 \\ 0 & 0 & 0 & 0 & 0 & 0.1 & 0.4 & -1.0 & 0.2 & 0.3 \\ 0 & 0 & 0 & 0 & 0 & 0.2 & 0.3 & 0.1 & -0.9 & 0.3 \\ 0 & 0 & 0 & 0 & 0 & 0.2 & 0.4 & 0.3 & 0.2 & -1.1 \end{pmatrix},$$

$$G^2 = \begin{pmatrix} -0.8 & 0.2 & 0.1 & 0.3 & 0.1 & 0.1 & 0 & 0 & 0 & 0 \\ 0.2 & -0.9 & 0.2 & 0.2 & 0.3 & 0 & 0 & 0 & 0 & 0 \\ 0.2 & 0.1 & -0.6 & 0.1 & 0.2 & 0 & 0 & 0 & 0 & 0 \\ 0.1 & 0.2 & 0.1 & -0.6 & 0.2 & 0 & 0 & 0 & 0 & 0 \\ 0.3 & 0.2 & 0.1 & 0.4 & -1.0 & 0 & 0 & & & \\ 0 & 0 & 0 & 0 & 0 & -0.5 & 0.1 & 0.1 & 0.2 & 0.1 \\ 0 & 0 & 0 & 0 & 0 & 0.1 & -0.8 & 0.2 & 0.3 & 0.2 \\ 0 & 0 & 0 & 0 & 0 & 0.1 & 0.2 & -0.8 & 0.3 & 0.2 \\ 0 & 0 & 0 & 0 & 0 & 0.2 & 0.3 & 0.4 & -1.2 & 0.3 \\ 0 & 0 & 0 & 0 & 0 & 0.3 & 0.2 & 0.1 & 0.2 & -0.8 \end{pmatrix},$$

$$G^3 = \begin{pmatrix} -0.9 & 0.2 & 0.1 & 0.3 & 0.1 & 0.2 & 0 & 0 & 0 & 0 \\ 0.1 & -0.8 & 0.2 & 0.2 & 0.3 & 0 & 0 & 0 & 0 & 0 \\ 0.2 & 0.1 & -0.8 & 0.1 & 0.4 & 0 & 0 & 0 & 0 & 0 \\ 0.1 & 0.2 & 0.3 & -0.7 & 0.1 & 0 & 0 & 0 & 0 & 0 \\ 0.3 & 0.1 & 0.2 & 0.3 & -0.9 & 0 & 0 & 0 & 0 & 0 \\ 0 & 0 & 0 & 0 & 0 & -0.6 & 0.1 & 0.2 & 0.2 & 0.1 \\ 0 & 0 & 0 & 0 & 0 & 0.1 & -0.9 & 0.3 & 0.3 & 0.2 \\ 0 & 0 & 0 & 0 & 0 & 0.2 & 0.2 & -0.7 & 0.1 & 0.2 \\ 0 & 0 & 0 & 0 & 0 & 0.1 & 0.2 & 0.2 & -0.6 & 0.1 \\ 0 & 0 & 0 & 0 & 0 & 0.2 & 0.1 & 0.1 & 0.1 & -0.5 \end{pmatrix}.$$

Obviously, $f_l(\cdot)(l = 1, 2, 3)$ satisfies the Lipschitz condition with $\rho_l = 0.5$, and $\hat{s}(t) = (0, 0, 0)^T \in \mathbb{R}^3$ is a solution of an isolated node of network (8.53). We select the nodes 1–7 as pinned nodes.

Case 1: Take $h_1^1 = 0.9, h_2^1 = 0.8, h_3^1 = 0.7, h_4^1 = 0.6, h_5^1 = 0.7, h_6^1 = 0.8, h_7^1 = 0.9, h_1^2 = 0.8, h_2^2 = 0.6, h_3^2 = 0.8, h_4^2 = 0.9, h_5^2 = 0.7, h_6^2 = 0.9, h_7^2 = 0.6, h_1^3 = 0.9, h_2^3 = 0.7, h_3^3 = 0.8, h_4^3 = 0.6, h_5^3 = 0.8, h_6^3 = 0.9$, and $h_7^3 = 0.8$.

Set $\alpha_1 = 0.1$, and $\alpha_2 = 0.2$. Using the MATLAB YALMIP Toolbox, we can get the following matrices Ψ_1 and Ψ_2 satisfying (8.9),

$$\Psi_1 = \text{diag}(0.7480, 0.8975, 0.8918, 0.9485, 0.7620),$$

$$\Psi_2 = \text{diag}(0.4496, 0.4131, 0.7383, 0.7287, 0.8053).$$

According to Theorem 8.3, the network (8.53) via pinning control achieves synchronization. The simulation results are displayed in Fig. 8.1.

Moreover, we can easily verify (8.19), (8.22) and (8.23) are satisfied if $\alpha_1 = 0.0815, \alpha_2 = 0.0810, b_1 = 0.1994, b_2 = 0.1762, b_3 = 0.1607, \Psi_1 = \text{diag}\ (3.6791, 4.3686, 4.3919, 4.5358, 3.6962), \Psi_2 = \text{diag}(4.4108, 4.1467, 6.7499, 6.0038, 6.6818)$. It follows from Theorem 8.7 that the network (8.53) under pinning control and adaptive laws (8.20) achieves synchronization. The simulation results are displayed in Figs. 8.2 and 8.3.

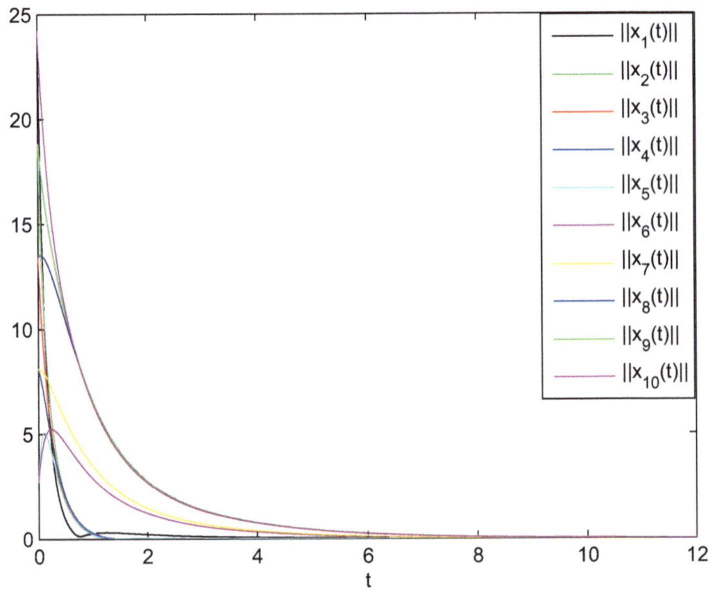

Fig. 8.1 The change processes of $\|x_i(t)\|, i = 1, 2, \cdots, 10$

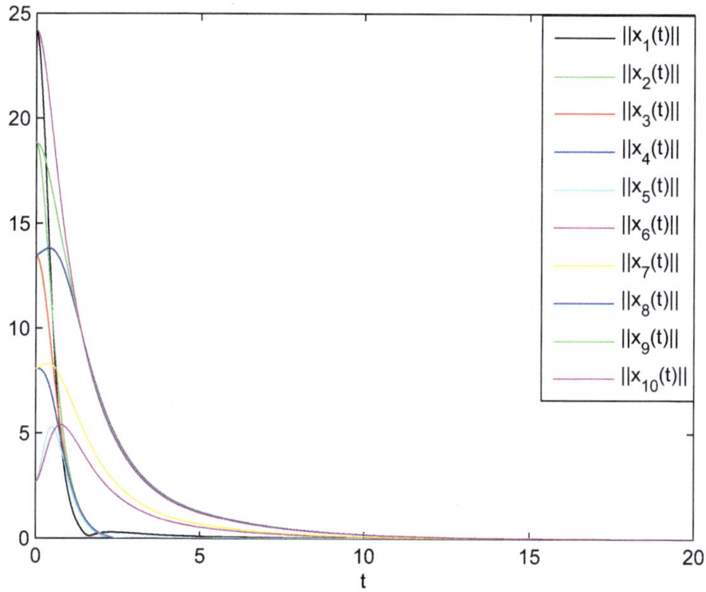

Fig. 8.2 The change processes of $\|x_i(t)\|$, $i = 1, 2, \cdots, 10$

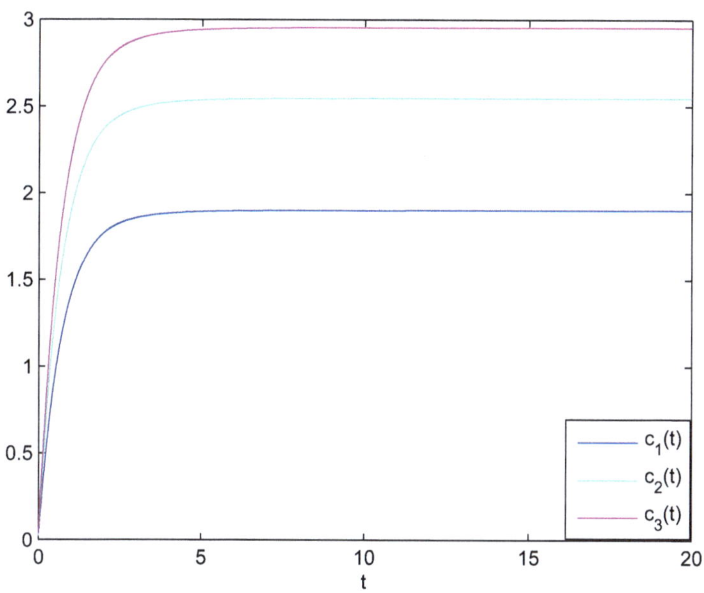

Fig. 8.3 Adaptive coupling strengths [$c_1(0) = 0.02$, $c_2(0) = 0.03$, $c_3(0) = 0.04$]

Case 2: Obviously, the condition (8.29) is satisfied if we take

$$\begin{aligned}
\Psi_1 &= \mathrm{diag}(0.7480, 0.8975, 0.8918, 0.9485, 0.7620),\\
\Psi_2 &= \mathrm{diag}(0.4496, 0.4131, 0.7383, 0.7287, 0.8053),\\
\hat{\kappa}_1^1 &= \mathrm{diag}(0.9, 0.8, 0.7, 0.6, 0.7),\\
\hat{\kappa}_2^1 &= \mathrm{diag}(0.8, 0.9, 0, 0, 0),\ \alpha_1 = 0.1,\ \alpha_2 = 0.2,\\
\hat{\kappa}_1^2 &= \mathrm{diag}(0.8, 0.6, 0.8, 0.9, 0.7),\\
\hat{\kappa}_2^2 &= \mathrm{diag}(0.9, 0.6, 0, 0, 0),\\
\hat{\kappa}_1^3 &= \mathrm{diag}(0.9, 0.7, 0.8, 0.6, 0.8),\\
\hat{\kappa}_2^3 &= \mathrm{diag}(0.9, 0.8, 0, 0, 0).
\end{aligned}$$

By Theorem 8.8, the network (8.53) can achieve synchronization under the pinning adaptive controllers. The simulation results are displayed in Figs. 8.4 and 8.5.

Example 8.14 Consider the following complex network with multi-weights and undirected topology:

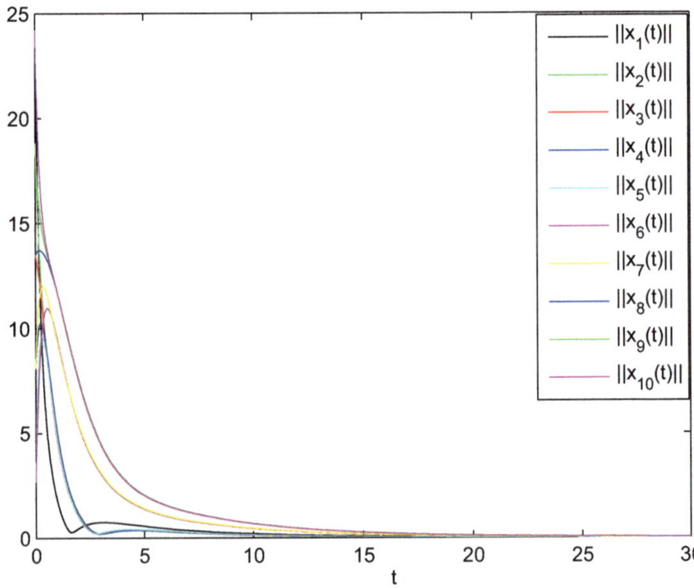

Fig. 8.4 The change processes of $\|x_i(t)\|, i = 1, 2, \cdots, 10$

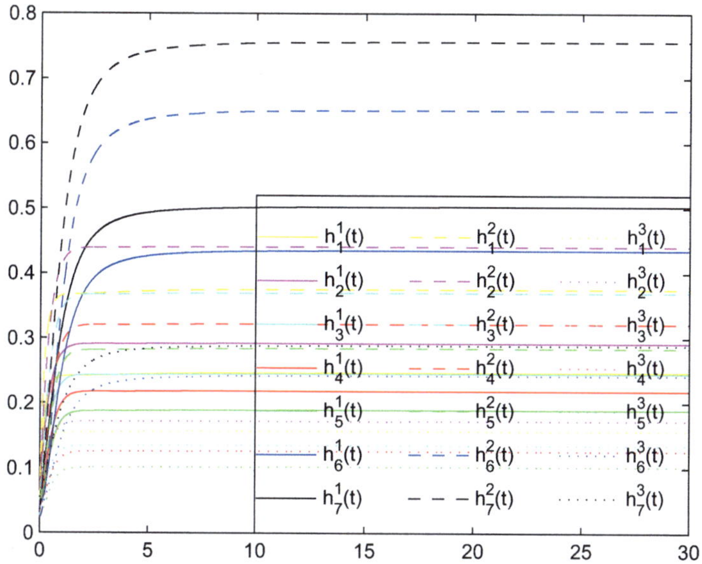

Fig. 8.5 The change processes of $h_i^r(t), r = 1, 2, 3, i = 1, 2, \cdots, 7$

$$\dot{x}_i(t) = f(x_i(t)) + c_1 \sum_{j=1}^{10} G_{ij}^1 \Gamma_1 x_j(t) + c_2 \sum_{j=1}^{10} G_{ij}^2 \Gamma_2 x_j(t) + c_3 \sum_{j=1}^{10} G_{ij}^3 \Gamma_3 x_j(t), \quad i = 1, 2, \cdots, 10, \tag{8.54}$$

where $f_l(\xi) = \frac{|\xi+1|-|\xi-1|}{4}, l = 1, 2, 3, \Gamma_1 = \mathrm{diag}(0.5, 0.8, 0.6), \Gamma_2 = \mathrm{diag}(0.7, 0.6, 0.5), \Gamma_3 = \mathrm{diag}(0.5, 0.7, 0.8), c_1 = 4, c_2 = 6, c_3 = 8$, the matrices $G^1 = (G_{ij}^1)_{10\times 10}$, $G^2 = (G_{ij}^2)_{10\times 10}$ and $G^3 = (G_{ij}^3)_{10\times 10}$ are chosen as, respectively

$$G^1 = \begin{pmatrix} -0.8 & 0.2 & 0.3 & 0.1 & 0.2 & 0 & 0 & 0 & 0 & 0 \\ 0.2 & -1.2 & 0.3 & 0.3 & 0.4 & 0 & 0 & 0 & 0 & 0 \\ 0.3 & 0.3 & -1.2 & 0.1 & 0.5 & 0 & 0 & 0 & 0 & 0 \\ 0.1 & 0.3 & 0.1 & -0.7 & 0.2 & 0 & 0 & 0 & 0 & 0 \\ 0.2 & 0.4 & 0.5 & 0.2 & -1.3 & 0 & 0 & 0 & 0 & 0 \\ 0 & 0 & 0 & 0 & 0 & -0.8 & 0.2 & 0.3 & 0.2 & 0.1 \\ 0 & 0 & 0 & 0 & 0 & 0.2 & -1.1 & 0.4 & 0.3 & 0.2 \\ 0 & 0 & 0 & 0 & 0 & 0.3 & 0.4 & -1.0 & 0.2 & 0.1 \\ 0 & 0 & 0 & 0 & 0 & 0.2 & 0.3 & 0.2 & -0.9 & 0.2 \\ 0 & 0 & 0 & 0 & 0 & 0.1 & 0.2 & 0.1 & 0.2 & -0.6 \end{pmatrix},$$

$$G^2 = \begin{pmatrix} -1.0 & 0.1 & 0.4 & 0.3 & 0.2 & 0 & 0 & 0 & 0 & 0 \\ 0.1 & -0.9 & 0.2 & 0.4 & 0.2 & 0 & 0 & 0 & 0 & 0 \\ 0.4 & 0.2 & -1.2 & 0.1 & 0.5 & 0 & 0 & 0 & 0 & 0 \\ 0.3 & 0.4 & 0.1 & -1.0 & 0.2 & 0 & 0 & 0 & 0 & 0 \\ 0.2 & 0.2 & 0.5 & 0.2 & -1.1 & 0 & 0 & 0 & 0 & 0 \\ 0 & 0 & 0 & 0 & 0 & -0.6 & 0.2 & 0.1 & 0.2 & 0.1 \\ 0 & 0 & 0 & 0 & 0 & 0.2 & -0.9 & 0.3 & 0.2 & 0.2 \\ 0 & 0 & 0 & 0 & 0 & 0.1 & 0.3 & -0.7 & 0.2 & 0.1 \\ 0 & 0 & 0 & 0 & 0 & 0.2 & 0.2 & 0.2 & -0.9 & 0.3 \\ 0 & 0 & 0 & 0 & 0 & 0.1 & 0.2 & 0.1 & 0.3 & -0.7 \end{pmatrix},$$

$$G^3 = \begin{pmatrix} -1.0 & 0.3 & 0.1 & 0.2 & 0.4 & 0 & 0 & 0 & 0 & 0 \\ 0.3 & -0.9 & 0.1 & 0.3 & 0.2 & 0 & 0 & 0 & 0 & 0 \\ 0.1 & 0.1 & -0.8 & 0.4 & 0.2 & 0 & 0 & 0 & 0 & 0 \\ 0.2 & 0.3 & 0.4 & -1.1 & 0.2 & 0 & 0 & 0 & 0 & 0 \\ 0.4 & 0.2 & 0.2 & 0.2 & -1.0 & 0 & 0 & 0 & 0 & 0 \\ 0 & 0 & 0 & 0 & 0 & -0.9 & 0.2 & 0.4 & 0.2 & 0.1 \\ 0 & 0 & 0 & 0 & 0 & 0.2 & -1.0 & 0.3 & 0.3 & 0.2 \\ 0 & 0 & 0 & 0 & 0 & 0.4 & 0.3 & -1.1 & 0.2 & 0.2 \\ 0 & 0 & 0 & 0 & 0 & 0.2 & 0.3 & 0.2 & -1.0 & 0.3 \\ 0 & 0 & 0 & 0 & 0 & 0.1 & 0.2 & 0.2 & 0.3 & -0.8 \end{pmatrix}.$$

Obviously, $f_l(\cdot)(l = 1, 2, 3)$ satisfies the Lipschitz condition with $\rho_l = 0.5$, and $\hat{s}(t) = (0, 0, 0)^T \in \mathbb{R}^3$ is a solution of an isolated node of network (8.54). We select the nodes 1–7 as pinned nodes.

Case 1: Take $h_1^1 = 0.8, h_2^1 = 0.7, h_3^1 = 0.6, h_4^1 = 0.8, h_5^1 = 0.7, h_6^1 = 0.8, h_7^1 = 0.9, h_1^2 = 0.8, h_2^2 = 0.7, h_3^2 = 0.6, h_4^2 = 0.7, h_5^2 = 0.9, h_6^2 = 0.8, h_7^2 = 0.6, h_1^3 = 0.7, h_2^3 = 0.8, h_3^3 = 0.9, h_4^3 = 0.7, h_5^3 = 0.9, h_6^3 = 0.9, h_7^3 = 0.7$.

By employing the MATLAB YALMIP Toolbox, we can get the following matrices Ψ_1 and Ψ_2 satisfying (8.38),

$$\Psi_1 = \text{diag}(0.3715, 0.3836, 0.3657, 0.3613, 0.3297),$$
$$\Psi_2 = \text{diag}(0.3816, 0.3771, 0.5582, 0.5898, 0.7580).$$

By Theorem 8.9, the network (8.54) via pinning control achieves synchronization. The simulation results are displayed in Fig. 8.6.

Moreover, we can easily verify that (8.45) is satisfied if

$$\Psi_1 = \text{diag}(3.5861, 3.4814, 3.3537, 3.6374, 3.0370),$$
$$\Psi_2 = \text{diag}(3.8059, 3.3797, 5.6382, 5.8241, 7.7071).$$

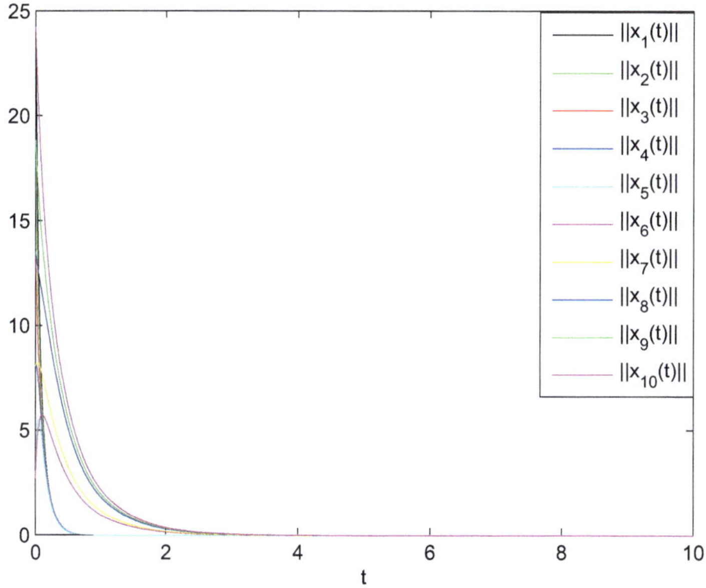

Fig. 8.6 The change processes of $\|x_i(t)\|, i = 1, 2, \cdots, 10$

From Theorem 8.11, the network (8.54) under pinning control and adaptive laws (8.46) achieves synchronization. The simulation results are displayed in Figs. 8.7 and 8.8.

Case 2: The following parameters can be found to satisfy (8.51):

$$\Psi_1 = \text{diag}(0.0277, 0.0274, 0.0267, 0.0276, 0.0239),$$

$$\Psi_2 = \text{diag}(0.0290, 0.0267, 0.0436, 0.0437, 0.0588),$$

$$\hat{\kappa}_1^1 = \text{diag}(0.8, 0.7, 0.6, 0.8, 0.7),$$

$$\hat{\kappa}_2^1 = \text{diag}(0.8, 0.9, 0, 0, 0),$$

$$\hat{\kappa}_1^2 = \text{diag}(0.8, 0.7, 0.6, 0.7, 0.9),$$

$$\hat{\kappa}_2^2 = \text{diag}(0.8, 0.6, 0, 0, 0),$$

$$\hat{\kappa}_1^3 = \text{diag}(0.7, 0.8, 0.9, 0.7, 0.9),$$

$$\hat{\kappa}_2^3 = \text{diag}(0.9, 0.7, 0, 0, 0).$$

According to Theorem 8.12, the network (8.54) achieves synchronization under the pinning adaptive controllers. The simulation results are displayed in Figs. 8.9 and 8.10.

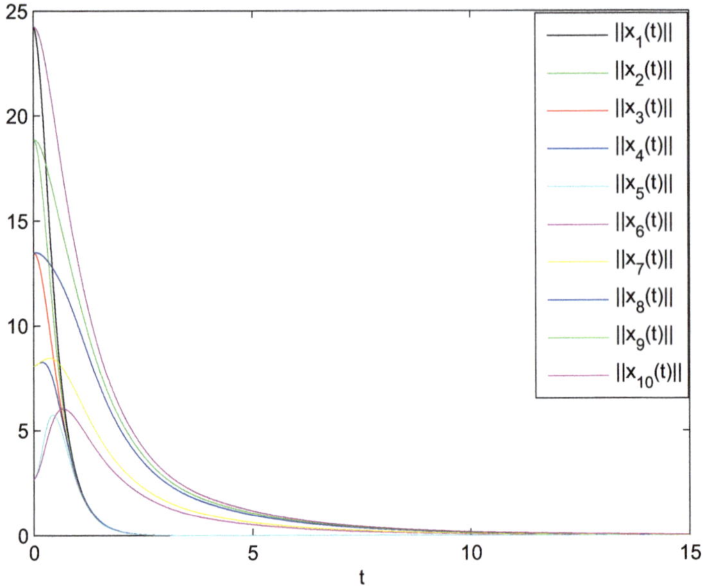

Fig. 8.7 The change processes of $\|x_i(t)\|, i = 1, 2, \cdots, 10$

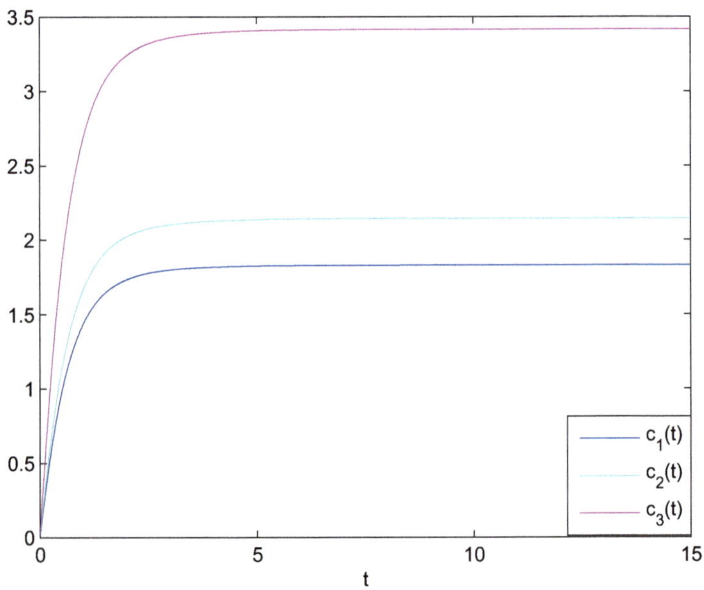

Fig. 8.8 Adaptive coupling strengths $[c_1(0) = 0.02, c_2(0) = 0.03, c_3(0) = 0.04]$

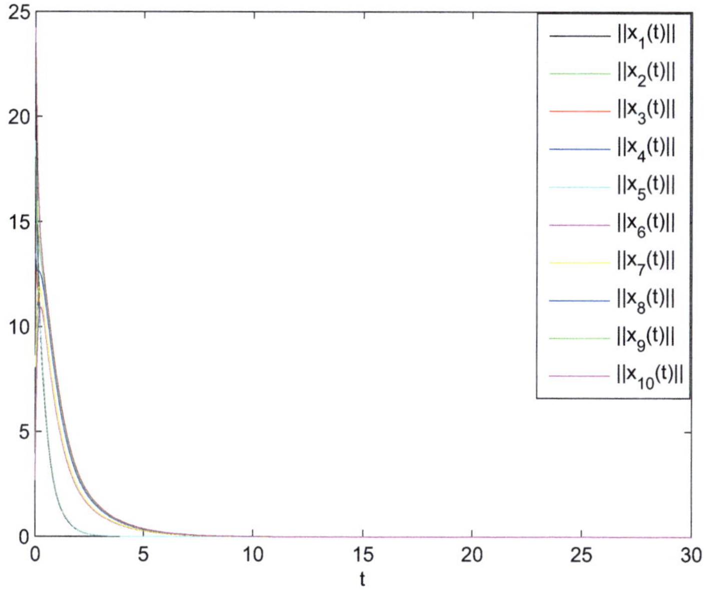

Fig. 8.9 The change processes of $\|x_i(t)\|, i = 1, 2, \cdots, 10$

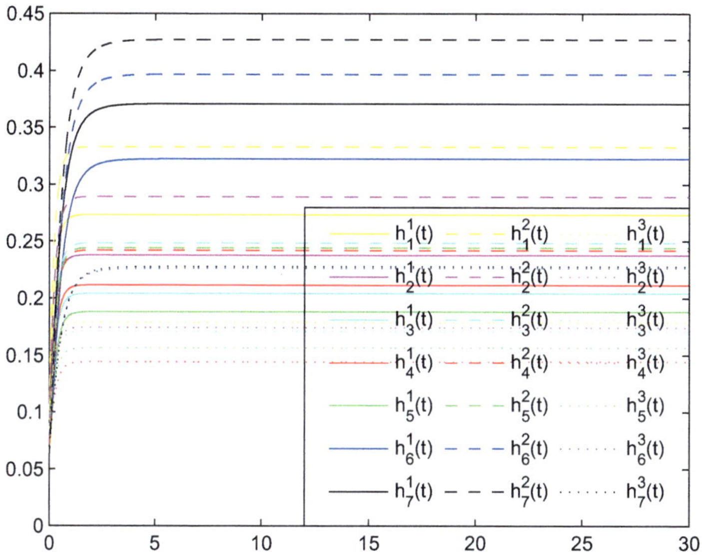

Fig. 8.10 The change processes of $h_i^r(t), r = 1, 2, 3, i = 1, 2, \cdots, 7$

8.4 Conclusion

In this chapter, the pinning synchronization for directed and undirected CDNs with multi-weights have been discussed, respectively. On one hand, some synchronization criteria have been established for the multi-weighted complex network with directed topology, and several adaptive laws for tuning coupling strengths and feedback gains have been presented. On the other hand, some similar results for undirected complex dynamical network with multi-weights also have been derived. Finally, illustrative examples have been given to demonstrate the correctness of these obtained results. In future work, we shall study the pinning synchronization problem of public traffic network with multi-weights, and analyze the effects of different coupling weights on synchronization.

Chapter 9
Analysis and Pinning Control for Output Synchronization and $\mathcal{H}_\infty$ Output Synchronization of Multi-weighted Complex Networks

9.1 Introduction

In recent years, the output synchronization of complex networks has attracted great attention of researchers, and a large number of criteria have been derived to guarantee output synchronization for different complex network models [24–38]. In [26], the authors took the output synchronization into consideration for an impulsive complex delayed dynamical network, and some criteria were acquired to ensure global exponential output synchronization of the above-mentioned network by means of impulsive delay differential inequalities. By exploiting the relationship between output synchronization and output strict passivity, Wang et al. [29] presented two output synchronization criteria for the complex networks with adaptive and fixed coupling strengths. In [30], the authors considered a kind of complex networks that not only has time-varying delay but also has output coupling, and discussed the output synchronization of this network model. Furthermore, external disturbances are ubiquitous and may disrupt output synchronization in complex networks. Therefore, it is crucial to consider external disturbances in complex networks and investigate their $\mathcal{H}_\infty$ output synchronization. To the best of our knowledge, very few results on this topic have been reported.

However, most of the existing results on output synchronization and $\mathcal{H}_\infty$ output synchronization are all based on network models with a single weight. Actually, many real-world networks, such as transportation networks, social networks, communication networks, and so on, are coupled by multiple coupling forms. Therefore, the investigation of multi-weighted complex networks is challenging and interesting. Nevertheless, the synchronization for multi-weighted complex networks have seldom been considered [39–41]. In [40], the authors derived a criterion about global synchronization of the multi-weighted public traffic roads networks on the basis of Lyapunov stability theory. Zhao et al. [41] studied the adaptive synchronization for a multiple delayed complex network with uncertain inner coupling matrix, and a synchronization criterion was presented for such network

J.-L. Wang et al., *Analysis and Control of Output Synchronization for Complex Dynamical Networks*, https://doi.org/10.1007/978-981-13-1352-3_9

model by employing the designed adaptive feedback controllers. Unfortunately, the output synchronization and $\mathcal{H}_\infty$ output synchronization of multi-weighted complex networks have not yet been investigated.

As is well known, complex networks usually find it hard to achieve synchronization by themselves. Thus, it is necessary to adopt some control schemes for realizing synchronization in complex networks. From a practical perspective, it is very tough to control all nodes or edges in a large-scale network. Consequently, many researchers have developed nodes-based pinning control strategies for synchronization and $\mathcal{H}_\infty$ synchronization of single weighted complex networks [9, 103, 163, 195, 197–206]. In [197], Guo et al. discussed the pinning adaptive cluster synchronization of the complex network that has Lurie-type nonlinear dynamics. By employing the pinning adaptive strategy, Ahmed et al. studied the exponential synchronization problem for a complex delayed network in [198]. To achieve synchronization in linearly coupled complex networks, Liu and Chen [201] designed an adaptive strategy for the pinning control gain. In [204], Xu et al. considered the global $\mathcal{H}_\infty$ pinning synchronization for a kind of complex networks with aperiodic samplings. In [205], by exploiting pinning control, the author discussed exponential output synchronization for a complex delayed network. Moreover, very few researchers have also considered synchronization and $\mathcal{H}_\infty$ synchronization problems for complex networks by using edges-based pinning control schemes [11, 88, 89, 148]. Yu et al. [11] investigated the synchronization of complex networks by adjusting a small part of coupling weights, and several criteria were obtained to ensure network synchronization by utilizing Lyapunov functionals and some inequality techniques. In [148], the authors considered the synchronization of fractional-order complex networks by use of the edges-based pinning control strategy. To guarantee $\mathcal{H}_\infty$ synchronization for coupled neural networks with reaction-diffusion terms and hybrid coupling, Wang and Wu [89] developed some adaptive strategies to update the coupling weights. Regrettably, the nodes-based and edges-based pinning control for output synchronization and $\mathcal{H}_\infty$ output synchronization of multi-weighted complex networks have not yet been investigated.

Motivated by the above-mentioned discussions, the output synchronization and $\mathcal{H}_\infty$ output synchronization of complex networks with multi-weights will be discussed in this chapter. First, we present two sufficient conditions for output synchronization and $\mathcal{H}_\infty$ output synchronization of multi-weighted complex networks using Lyapunov functionals and Barbalat's Lemma. Second, a nodes-based pinning control strategy is given to ensure output synchronization and $\mathcal{H}_\infty$ output synchronization for multi-weighted complex networks. Third, for the sake of ensuring output synchronization and $\mathcal{H}_\infty$ output synchronization of multi-weighted complex networks, a edges-based pinning control strategy is also developed.

9.2 Output Synchronization and Pinning Control of Complex Networks with Multi-weights

9.2.1 Network Model

The multi-weighted complex network to be investigated in this section is given as follows:

$$\begin{aligned}\dot{x}_i(t) &= f(x_i(t)) + a_1 \sum_{j=1}^{N} G_{ij}^1 \Gamma_1 x_j(t) + a_2 \sum_{j=1}^{N} G_{ij}^2 \Gamma_2 x_j(t) \\ &\quad + \cdots + a_\sigma \sum_{j=1}^{N} G_{ij}^\sigma \Gamma_\sigma x_j(t), \\ y_i(t) &= C x_i(t), \; i = 1, 2, \cdots, N, \end{aligned} \tag{9.1}$$

where $f(\cdot) : \mathbb{R}^n \to \mathbb{R}^n$ is a continuously differentiable vector function; $x_i(t) = (x_{i1}(t), x_{i2}(t), \cdots, x_{in}(t))^T \in \mathbb{R}^n$ is the state vector of the ith node; $y_i(t) = (y_{i1}(t), y_{i2}(t), \cdots, y_{ip}(t))^T \in \mathbb{R}^p$ is the output vector of the ith node; $C \in \mathbb{R}^{p\times n}$ is a known matrix; a_r $(r = 1, 2, \cdots, \sigma)$ denotes the coupling strength of the rth coupling form and is a positive real number; $\Gamma_r \in \mathbb{R}^{n\times n} > 0$ $(r = 1, 2, \cdots, \sigma)$ is the inner coupling matrix for the rth coupling form; $G^r = (G_{ij}^r)_{N\times N}$ $(r = 1, 2, \cdots, \sigma)$ is the coupling configuration matrix that denotes coupling weights in the rth coupling form, where G_{ij}^r is defined as follows: if nodes i and j are connected, then $G_{ij}^r = G_{ji}^r > 0$; or else, $G_{ij}^r = G_{ji}^r = 0 (i \neq j)$, the diagonal elements of matrix G^r satisfy

$$G_{ii}^r = -\sum_{\substack{j=1 \\ j\neq i}}^{N} G_{ij}^r, \, i = 1, 2, \cdots, N.$$

In this section, the complex network (9.1) is always required to be connected, and it has the same topology structure for different coupling forms. That is, $G_{ij}^r = 0$ $(r \in \{1, 2, \cdots, \sigma\})$ if and only if $G_{ij}^\varsigma = 0$ for $\varsigma = 1, 2, \cdots, r-1, r+1, \cdots, \sigma$.

Remark 9.1 As far as we know, many networks in the real world can be described by multi-weighted complex network models, such as transportation networks, communication networks, social networks, and so on. For all we know, the synchronization for multi-weighted complex networks have seldom been considered [39–41]. Specifically, the output synchronization for multi-weighted complex networks has never been discussed.

To derive our results, an assumption is introduced as follows:

(A1). There exists a matrix $\mathbb{R}^{n\times n} \ni Q = \mathrm{diag}(q_1, q_2, \cdots, q_n) > 0$ and a matrix $\mathbb{R}^{n\times n} \ni \Xi = \mathrm{diag}(\xi_1, \xi_2, \cdots, \xi_n)$ such that $f(\cdot)$ satisfies the following inequality:

$$(\iota_1 - \iota_2)^T Q\big[f(\iota_1) - f(\iota_2) - \Xi(\iota_1 - \iota_2)\big] \leqslant -\eta(\iota_1 - \iota_2)^T(\iota_1 - \iota_2)$$

for some $0 < \eta \in \mathbb{R}$ and all $\iota_1, \iota_2 \in \mathbb{R}^n$.

Remark 9.2 Throughout this chapter, we always assume that the function $f(\cdot)$ satisfies Lipschitz condition, for the sake of ensuring the uniqueness and existence of the solution of a complex network (9.1). In fact, some authors have proved that (A1) obviously holds under the condition that $f(\cdot)$ satisfies Lipschitz condition [112].

Definition 9.3 (see [30]) The complex network (9.1) is output synchronized if

$$\lim_{t\to+\infty}\left\| y_i(t) - \frac{1}{N}\sum_{k=1}^{N} y_k(t)\right\|_2 = 0 \text{ for all } i = 1, 2, \cdots, N.$$

9.2.2 Output Synchronization

Letting $\bar{x}(t) = \frac{1}{N}\sum_{i=1}^{N} x_i(t)$, one has

$$\begin{aligned}\dot{\bar{x}}(t) &= \frac{1}{N}\sum_{i=1}^{N} f(x_i(t)) + \frac{a_1}{N}\sum_{j=1}^{N}\left(\sum_{i=1}^{N} G_{ij}^1\right)\Gamma_1 x_j(t) \\ &\quad + \frac{a_2}{N}\sum_{j=1}^{N}\left(\sum_{i=1}^{N} G_{ij}^2\right)\Gamma_2 x_j(t) + \cdots \\ &\quad + \frac{a_\sigma}{N}\sum_{j=1}^{N}\left(\sum_{i=1}^{N} G_{ij}^\sigma\right)\Gamma_\sigma x_j(t) \\ &= \frac{1}{N}\sum_{i=1}^{N} f(x_i(t)).\end{aligned} \tag{9.2}$$

Define $e_i(t) = x_i(t) - \bar{x}(t)$, $i = 1, 2, \cdots, N$. Then, we have

$$\begin{aligned}\dot{e}_i(t) &= f(x_i(t)) - \frac{1}{N}\sum_{k=1}^{N} f(x_k(t)) + a_1\sum_{j=1}^{N} G_{ij}^1\Gamma_1 e_j(t) \\ &\quad + a_2\sum_{j=1}^{N} G_{ij}^2\Gamma_2 e_j(t) + \cdots + a_\sigma\sum_{j=1}^{N} G_{ij}^\sigma\Gamma_\sigma e_j(t),\end{aligned}$$

$$y_i(t) = Ce_i(t) + C\bar{x}(t), \tag{9.3}$$

where $i = 1, 2, \cdots, N$.

Obviously, the network (9.3) realizes output synchronization if and only if $\lim_{t\to+\infty} y_i^*(t) = 0, i = 1, 2, \cdots, N$, hold in system (9.4):

$$\begin{aligned}
\dot{e}_i(t) &= f(x_i(t)) - \frac{1}{N}\sum_{k=1}^{N} f(x_k(t)) + a_1 \sum_{j=1}^{N} G_{ij}^1 \Gamma_1 e_j(t) \\
&\quad + a_2 \sum_{j=1}^{N} G_{ij}^2 \Gamma_2 e_j(t) + \cdots + a_\sigma \sum_{j=1}^{N} G_{ij}^\sigma \Gamma_\sigma e_j(t), \\
y_i^*(t) &= Ce_i(t),
\end{aligned} \tag{9.4}$$

where $y_i^*(t) = y_i(t) - C\bar{x}(t), i = 1, 2, \cdots, N$.

On the other hand, we can easily find the non-singular matrices $M_1 \in \mathbb{R}^{p\times p}$ and $M_2 \in \mathbb{R}^{n\times n}$ such that

$$C = M_1 \begin{pmatrix} I_b & 0 \\ 0 & 0 \end{pmatrix} M_2.$$

Then, one gets

$$\begin{aligned}
\dot{e}_i(t) &= f(x_i(t)) - \frac{1}{N}\sum_{k=1}^{N} f(x_k(t)) + a_1 \sum_{j=1}^{N} G_{ij}^1 \Gamma_1 e_j(t) \\
&\quad + a_2 \sum_{j=1}^{N} G_{ij}^2 \Gamma_2 e_j(t) + \cdots + a_\sigma \sum_{j=1}^{N} G_{ij}^\sigma \Gamma_\sigma e_j(t), \\
\hat{y}_i^*(t) &= \hat{C}e_i(t),
\end{aligned} \tag{9.5}$$

where $\hat{y}_i^*(t) = (\hat{y}_{i1}^*(t), \hat{y}_{i2}^*(t), \cdots, \hat{y}_{ib}^*(t))^T \in \mathbb{R}^b, \hat{C} = (I_b \quad 0)_{b\times n} M_2, i = 1, 2, \cdots, N$. Apparently, we can easily conclude that $\lim_{t\to+\infty} y_i^*(t) = 0$ is equivalent to $\lim_{t\to+\infty} \hat{y}_i^*(t) = 0$.

Theorem 9.4 *If there exists a positive real number ξ such that*

$$2I_N \otimes (Q\Xi - \eta I_n) + \sum_{r=1}^{\sigma} a_r G^r \otimes (Q\Gamma_r + \Gamma_r Q) + W \leqslant 0, \tag{9.6}$$

where $W = \xi I_N \otimes (\hat{C}^T \hat{C})$, then the network (9.1) *realizes output synchronization.*

Proof For the network (9.5), the Lyapunov functional is chosen as follows:

$$V_1(t) = \sum_{i=1}^{N} e_i^T(t) Q e_i(t). \tag{9.7}$$

Then, we acquire

$$\begin{aligned}\dot{V}_1(t) = 2\sum_{i=1}^{N} e_i^T(t) Q \Big[f(x_i(t)) - \frac{1}{N}\sum_{k=1}^{N} f(x_k(t)) \\ + f(\bar{x}(t)) - f(\bar{x}(t)) + a_1 \sum_{j=1}^{N} G_{ij}^1 \Gamma_1 e_j(t) \\ + a_2 \sum_{j=1}^{N} G_{ij}^2 \Gamma_2 e_j(t) + \cdots + a_\sigma \sum_{j=1}^{N} G_{ij}^\sigma \Gamma_\sigma e_j(t) \Big].\end{aligned}$$

Furthermore,

$$\begin{aligned}\sum_{i=1}^{N} e_i(t) &= \sum_{i=1}^{N} \left(x_i(t) - \frac{1}{N}\sum_{k=1}^{N} x_k(t) \right) \\ &= \sum_{i=1}^{N} x_i(t) - \sum_{k=1}^{N} x_k(t) \\ &= 0.\end{aligned} \tag{9.8}$$

From (9.8), one has

$$\sum_{i=1}^{N} e_i^T(t) Q \left(f(\bar{x}(t)) - \frac{1}{N}\sum_{k=1}^{N} f(x_k(t)) \right) = 0. \tag{9.9}$$

On the other hand,

$$\sum_{i=1}^{N} e_i^T(t) Q \big(f(x_i(t)) - f(\bar{x}(t)) \big) \leqslant \sum_{i=1}^{N} e_i^T(t) (Q\Xi - \eta I_n) e_i(t). \tag{9.10}$$

Thus,

$$\dot{V}_1(t) \leqslant \sum_{i=1}^{N} e_i^T(t) \Big[2(Q\Xi - \eta I_n) e_i(t) + 2a_1 \sum_{j=1}^{N} G_{ij}^1 Q \Gamma_1 e_j(t)$$

$$
\begin{aligned}
&+2a_2 \sum_{j=1}^{N} G_{ij}^2 Q\Gamma_2 e_j(t) + \cdots + 2a_\sigma \sum_{j=1}^{N} G_{ij}^\sigma Q\Gamma_\sigma e_j(t)\Big] \\
&= e^T(t)\Big[2I_N \otimes (Q\Xi - \eta I_n) + \sum_{r=1}^{\sigma} a_r G^r \otimes (Q\Gamma_r + \Gamma_r Q)\Big]e(t) \\
&\leqslant -\xi e^T(t)\big[I_N \otimes (\hat{C}^T\hat{C})\big]e(t) \\
&= -\xi \sum_{i=1}^{N} e_i^T(t)\hat{C}^T\hat{C}e_i(t), \qquad (9.11)
\end{aligned}
$$

where $e(t) = (e_1^T(t), e_2^T(t), \cdots, e_N^T(t))^T$.

According to (9.11), $V_1(t)$ is nonincreasing. Thus, $e_i(t), i = 1, 2, \cdots, N$ are bounded. Namely,

$$
\|e_i(t)\|_2 \leqslant \varphi_i \in \mathbb{R} > 0 \qquad (9.12)
$$

for any $t \in [0, +\infty)$.

In addition, we can easily get

$$
\lim_{t\to+\infty} V_1(t) = \delta \in \mathbb{R} \geqslant 0. \qquad (9.13)
$$

Let $\beta(t) = \sum_{i=1}^{N} e_i^T(t)\hat{C}^T\hat{C}e_i(t)$. From (9.11) and (9.13), one has

$$
\begin{aligned}
\lim_{t\to+\infty} \int_0^t \beta(s)ds &\leqslant -\frac{1}{\xi} \lim_{t\to+\infty} \int_0^t \dot{V}_1(s)ds \\
&= -\frac{1}{\xi}\big(\lim_{t\to+\infty} V_1(t) - V_1(0)\big) \\
&= -\frac{1}{\xi}\big(\delta - V_1(0)\big). \qquad (9.14)
\end{aligned}
$$

Obviously,

$$
\begin{aligned}
\dot{\beta}(t) = 2\sum_{i=1}^{N} e_i^T(t)\hat{C}^T\hat{C}\Big[f(x_i(t)) - \frac{1}{N}\sum_{k=1}^{N} f(x_k(t)) \\
+ f(\bar{x}(t)) - f(\bar{x}(t)) + a_1 \sum_{j=1}^{N} G_{ij}^1 \Gamma_1 e_j(t)
\end{aligned}
$$

$$+a_2 \sum_{j=1}^{N} G_{ij}^2 \Gamma_2 e_j(t) + \cdots + a_\sigma \sum_{j=1}^{N} G_{ij}^\sigma \Gamma_\sigma e_j(t)\Big]$$

$$= 2\sum_{i=1}^{N} e_i^T(t)\hat{C}^T\hat{C}\Big[f(x_i(t)) - f(\bar{x}(t))$$

$$+a_1 \sum_{j=1}^{N} G_{ij}^1 \Gamma_1 e_j(t) + a_2 \sum_{j=1}^{N} G_{ij}^2 \Gamma_2 e_j(t)$$

$$+ \cdots + a_\sigma \sum_{j=1}^{N} G_{ij}^\sigma \Gamma_\sigma e_j(t)\Big]. \tag{9.15}$$

Because $f(\cdot)$ fulfills the Lipschitz condition, there exists a positive real number α such that

$$\|f(x_i(t)) - f(\bar{x}(t))\|_2 \leqslant \alpha\|x_i(t) - \bar{x}(t)\|_2 = \alpha\|e_i(t)\|_2, \tag{9.16}$$

where $i = 1, 2, \cdots, N$.

By means of (9.12), (9.15) and (9.16), one obtains

$$|\dot{\beta}(t)| \leqslant 2\sum_{i=1}^{N} \Big|e_i^T(t)\hat{C}^T\hat{C}[f(x_i(t)) - f(\bar{x}(t))]\Big|$$

$$+2a_1 \sum_{i=1}^{N}\sum_{j=1}^{N} |G_{ij}^1| \Big|e_i^T(t)\hat{C}^T\hat{C}\Gamma_1 e_j(t)\Big|$$

$$+2a_2 \sum_{i=1}^{N}\sum_{j=1}^{N} |G_{ij}^2| \Big|e_i^T(t)\hat{C}^T\hat{C}\Gamma_2 e_j(t)\Big|$$

$$+ \cdots + 2a_\sigma \sum_{i=1}^{N}\sum_{j=1}^{N} |G_{ij}^\sigma| \Big|e_i^T(t)\hat{C}^T\hat{C}\Gamma_\sigma e_j(t)\Big|$$

$$\leqslant 2\sum_{i=1}^{N} \big\|\hat{C}^T\hat{C}e_i(t)\big\|_2 \big\|f(x_i(t)) - f(\bar{x}(t))\big\|_2$$

$$+2a_1 \sum_{i=1}^{N}\sum_{j=1}^{N} |G_{ij}^1| \big\|\hat{C}^T\hat{C}e_i(t)\big\|_2 \big\|\Gamma_1 e_j(t)\big\|_2$$

$$+2a_2 \sum_{i=1}^{N}\sum_{j=1}^{N} |G_{ij}^2| \big\|\hat{C}^T\hat{C}e_i(t)\big\|_2 \big\|\Gamma_2 e_j(t)\big\|_2$$

$$+\cdots+2a_\sigma\sum_{i=1}^{N}\sum_{j=1}^{N}|G_{ij}^{\sigma}|\big\|\hat{C}^T\hat{C}e_i(t)\big\|_2\big\|\Gamma_\sigma e_j(t)\big\|_2$$

$$\leqslant\sum_{i=1}^{N}\Big(\lambda_M^2(\hat{C}^T\hat{C})+\alpha^2\Big)\varphi_i^2$$

$$+a_1\sum_{i=1}^{N}\sum_{j=1}^{N}|G_{ij}^{1}|\Big(\lambda_M^2(\hat{C}^T\hat{C})\varphi_i^2+\lambda_M^2(\Gamma_1)\varphi_j^2\Big)$$

$$+a_2\sum_{i=1}^{N}\sum_{j=1}^{N}|G_{ij}^{2}|\Big(\lambda_M^2(\hat{C}^T\hat{C})\varphi_i^2+\lambda_M^2(\Gamma_2)\varphi_j^2\Big)$$

$$+\cdots+a_\sigma\sum_{i=1}^{N}\sum_{j=1}^{N}|G_{ij}^{\sigma}|\Big(\lambda_M^2(\hat{C}^T\hat{C})\varphi_i^2+\lambda_M^2(\Gamma_\sigma)\varphi_j^2\Big). \quad (9.17)$$

By (9.17), we can easily acquire that $\beta(t)$ is uniformly continuous. From Barbalat's Lemma (see [50]), we get

$$\lim_{t\to+\infty}\beta(t)=0. \quad (9.18)$$

Using (9.18), one has

$$\lim_{t\to+\infty}\|\hat{y}_i^*(t)\|_2=\lim_{t\to+\infty}\|\hat{C}e_i(t)\|_2=0, \quad (9.19)$$

where $i=1,2,\cdots,N$.

Consequently, we can deduce that the network (9.1) is output synchronized. □

9.2.3 *Nodes-Based Pinning Output Synchronization*

Let the first l $(1\leqslant l<N)$ nodes be pinned for realizing output synchronization in network (9.1). Subsequently, the controlled network is given as follows:

$$\dot{x}_i(t)=f(x_i(t))+a_1\sum_{j=1}^{N}G_{ij}^{1}\Gamma_1x_j(t)+a_2\sum_{j=1}^{N}G_{ij}^{2}\Gamma_2x_j(t)$$

$$+\cdots+a_\sigma\sum_{j=1}^{N}G_{ij}^{\sigma}\Gamma_\sigma x_j(t)+u_i(t),\ i=1,2,\cdots,l,$$

$$
\begin{aligned}
\dot{x}_i(t) &= f(x_i(t)) + a_1\sum_{j=1}^{N} G_{ij}^1 \Gamma_1 x_j(t) + a_2\sum_{j=1}^{N} G_{ij}^2 \Gamma_2 x_j(t) \\
&\quad + \cdots + a_\sigma \sum_{j=1}^{N} G_{ij}^\sigma \Gamma_\sigma x_j(t), \ i = l+1, l+2, \cdots, N, \\
y_i(t) &= C x_i(t), i = 1, 2, \cdots, N,
\end{aligned} \tag{9.20}
$$

where

$$
\begin{aligned}
u_i(t) &= -\sum_{r=1}^{\sigma} a_r h_i^r(t) \Gamma_r \left(x_i(t) - \frac{1}{N}\sum_{k=1}^{N} x_k(t) \right), \\
\dot{h}_i^r(t) &= \beta_i^r \left(x_i(t) - \frac{1}{N}\sum_{k=1}^{N} x_k(t) \right)^T \frac{Q\Gamma_r + \Gamma_r Q}{2} \\
&\quad \times \left(x_i(t) - \frac{1}{N}\sum_{k=1}^{N} x_k(t) \right),
\end{aligned} \tag{9.21}
$$

in which $Q\Gamma_r + \Gamma_r Q > 0, \mathbb{R} \ni \beta_i^r > 0, \mathbb{R} \ni h_i^r(0) > 0, i = 1, 2, \cdots, l, h_i^r(t) \equiv 0$ for $i = l+1, l+2, \cdots, N$.

Define $\bar{x}(t) = \frac{1}{N}\sum_{i=1}^{N} x_i(t)$. Then, one has

$$
\dot{\bar{x}}(t) = \frac{1}{N}\sum_{i=1}^{N} f(x_i(t)) + \frac{1}{N}\sum_{i=1}^{l} u_i(t).
$$

Let $e_i(t) = x_i(t) - \bar{x}(t)$. Similarly, we can easily deduce that the network (9.20) is output synchronized if and only if $\lim_{t\to+\infty} \hat{y}_i^*(t) = 0, i = 1, 2, \cdots, N$, hold in system (9.22):

$$
\begin{aligned}
\dot{e}_i(t) &= f(x_i(t)) - \frac{1}{N}\sum_{k=1}^{N} f(x_k(t)) - \frac{1}{N}\sum_{k=1}^{l} u_k(t) \\
&\quad + \sum_{r=1}^{\sigma}\sum_{j=1}^{N} a_r G_{ij}^r \Gamma_r e_j(t) - \sum_{r=1}^{\sigma} a_r h_i^r(t) \Gamma_r e_i(t) \\
\hat{y}_i^*(t) &= \hat{C} e_i(t), i = 1, 2, \cdots, N, \\
\dot{h}_i^r(t) &= \beta_i^r e_i^T(t) \frac{Q\Gamma_r + \Gamma_r Q}{2} e_i(t), i = 1, 2, \cdots, l,
\end{aligned} \tag{9.22}
$$

where $\hat{y}_i^*(t)$ and $\hat{C}$ denote the same meanings as these in (9.5).

Theorem 9.5 *If there exist matrices* $H^r = \operatorname{diag}(\hat{h}_1^r, \hat{h}_2^r, \cdots, \hat{h}_l^r, 0, \cdots, 0) \in \mathbb{R}^{N\times N}$, $r = 1, 2, \cdots, \sigma$, *and a positive real number* ξ *such that*

$$2I_N \otimes (Q\Xi - \eta I_n) + W_1 + \xi I_N \otimes \hat{C}^T \hat{C} \leqslant 0, \tag{9.23}$$

where $W_1 = \sum_{r=1}^{\sigma} a_r(G^r - H^r) \otimes (Q\Gamma_r + \Gamma_r Q)$, $\mathbb{R} \ni \hat{h}_i^r > 0$, *then the network* (9.20) *realizes output synchronization.*

Proof For the network (9.22), the Lyapunov functional is chosen as follows:

$$V_2(t) = \sum_{i=1}^{N} e_i^T(t) Q e_i(t) + \sum_{r=1}^{\sigma}\sum_{i=1}^{l} \frac{a_r}{\beta_i^r}(h_i^r(t) - \hat{h}_i^r)^2. \tag{9.24}$$

Then, one gets

$$\begin{aligned}\dot{V}_2(t) = {} & 2\sum_{i=1}^{N} e_i^T(t) Q\Big[f(x_i(t)) - \frac{1}{N}\sum_{k=1}^{N} f(x_k(t)) + f(\bar{x}(t)) - f(\bar{x}(t)) \\ & + \sum_{r=1}^{\sigma}\sum_{j=1}^{N} a_r G_{ij}^r \Gamma_r e_j(t) - \frac{1}{N}\sum_{k=1}^{l} u_k(t) - \sum_{r=1}^{\sigma} a_r h_i^r(t)\Gamma_r e_i(t)\Big] \\ & + \sum_{r=1}^{\sigma}\sum_{i=1}^{l} a_r(h_i^r(t) - \hat{h}_i^r) e_i^T(t)(Q\Gamma_r + \Gamma_r Q) e_i(t).\end{aligned} \tag{9.25}$$

As a result of $\sum_{i=1}^{N} e_i(t) = \sum_{i=1}^{N}(x_i(t) - \bar{x}(t)) = 0$, one has

$$\sum_{i=1}^{N} e_i^T(t) Q \left(\frac{1}{N}\sum_{k=1}^{l} u_k(t)\right) = 0. \tag{9.26}$$

Thus,

$$\begin{aligned}\dot{V}_2(t) \leqslant {} & 2\sum_{i=1}^{N} e_i^T(t)(Q\Xi - \eta I_n) e_i(t) + 2\sum_{r=1}^{\sigma}\sum_{i=1}^{N}\sum_{j=1}^{N} a_r G_{ij}^r e_i^T(t) Q\Gamma_r e_j(t) \\ & - \sum_{r=1}^{\sigma}\sum_{i=1}^{N} a_r \hat{h}_i^r e_i^T(t)(Q\Gamma_r + \Gamma_r Q) e_i(t) \\ = {} & e^T(t)\Big[2I_N \otimes (Q\Xi - \eta I_n) + \sum_{r=1}^{\sigma} a_r G^r \otimes (Q\Gamma_r + \Gamma_r Q)\end{aligned}$$

$$
\begin{aligned}
&-\sum_{r=1}^{\sigma} a_r H^r \otimes (Q\Gamma_r + \Gamma_r Q)\Big]e(t)\\
&\leqslant -\xi e^T(t)[I_N \otimes (\hat{C}^T\hat{C})]e(t)\\
&= -\xi \sum_{i=1}^{N} e_i^T(t)\hat{C}^T\hat{C}e_i(t),
\end{aligned}
\tag{9.27}
$$

where $e(t) = (e_1^T(t), e_2^T(t), \cdots, e_N^T(t))^T, \hat{h}_i^r = 0, i = l+1, l+2, \cdots, N$.

From (9.27), $V_2(t)$ is nonincreasing. Hence, $e_i(t)$ and $h_i^r(t), r = 1, 2, \cdots, \sigma, i = 1, 2, \cdots, N$, are bounded. Namely,

$$
\begin{aligned}
\|e_i(t)\|_2 &\leqslant \varphi_i \in \mathbb{R} > 0,\\
|h_i^r(t)| &\leqslant \varpi_i^r \in \mathbb{R} > 0
\end{aligned}
$$

for any $t \in [0, +\infty)$.

Then, by using similar method as in Theorem 9.4, one obtains

$$
\lim_{t\to+\infty} \|\hat{y}_i^*(t)\|_2 = \lim_{t\to+\infty} \|\hat{C}e_i(t)\|_2 = 0,
$$

where $i = 1, 2, \cdots, N$.

Consequently, we can conclude that the network (9.20) is output synchronized. □

9.2.4 *Edges-Based Pinning Output Synchronization*

Suppose $E \subseteq \{1, 2, \cdots, N\} \times \{1, 2, \cdots, N\}$ represents the set of undirected edges in the complex network (9.1), and network (9.1) is connected through $\hat{E} \subset E$. The complex network (9.1) with a portion of adaptive coupling weights is described by [108]

$$
\begin{aligned}
\dot{x}_i(t) &= f(x_i(t)) + a_1 \sum_{j=1}^{N} G_{ij}^1(t)\Gamma_1 x_j(t) + a_2 \sum_{j=1}^{N} G_{ij}^2(t)\Gamma_2 x_j(t) + \cdots\\
&\quad + a_\sigma \sum_{j=1}^{N} G_{ij}^\sigma(t)\Gamma_\sigma x_j(t),\\
y_i(t) &= Cx_i(t),
\end{aligned}
\tag{9.28}
$$

in which

$$\dot{G}^r_{ij}(t) = \beta^r_{ij}(x_i(t) - x_j(t))^T (Q\Gamma_r + \Gamma_r Q)(x_i(t) - x_j(t)), \ \text{if} \ (i, j) \in \hat{E}$$
$$\dot{G}^r_{ij}(t) = 0, \qquad \text{if} \ (i, j) \in E - \hat{E}, \tag{9.29}$$

where $i = 1, 2, \cdots, N, \mathbb{R} \ni \beta^r_{ij} = \beta^r_{ji} > 0, Q\Gamma_r + \Gamma_r Q > 0, r = 1, 2, \cdots, \sigma$.

Take $\bar{x}(t) = \frac{1}{N}\sum_{i=1}^N x_i(t)$ and $e_i(t) = x_i(t) - \bar{x}(t)$. Obviously, the complex network (9.28) is output synchronized under adaptive law (9.29) if and only if $\lim_{t\to+\infty} \hat{y}^*_i(t) = 0, i = 1, 2, \cdots, N$, hold in system (9.30):

$$\dot{e}_i(t) = f(x_i(t)) - \frac{1}{N}\sum_{k=1}^N f(x_k(t)) + \sum_{r=1}^\sigma \sum_{j=1}^N a_r G^r_{ij}(t)\Gamma_r e_j(t)$$
$$\hat{y}^*_i(t) = \hat{C}e_i(t), i = 1, 2, \cdots, N,$$
$$\dot{G}^r_{ij}(t) = \beta^r_{ij}(e_i(t) - e_j(t))^T (Q\Gamma_r + \Gamma_r Q)(e_i(t) - e_j(t)), \ \text{if} \ (i, j) \in \hat{E}$$
$$\dot{G}^r_{ij}(t) = 0, \qquad \text{if} \ (i, j) \in E - \hat{E}, \tag{9.30}$$

where $\hat{y}^*_i(t)$ and $\hat{C}$ have the same meanings as these in (9.5).

Denote

$$O^r_{ij} = \begin{cases} G^r_{ij}(0), & \text{if} \ (i, j) \in E - \hat{E}, \\ -\sum_{\substack{j=1 \\ j\neq i}}^N O^r_{ij}, & \text{if} \ i = j, \\ 0, & \text{otherwise.} \end{cases}$$

$$\Psi^r_{ij}(t) = \begin{cases} G^r_{ij}(t), & \text{if} \ (i, j) \in \hat{E}, \\ -\sum_{\substack{j=1 \\ j\neq i}}^N \Psi^r_{ij}(t), & \text{if} \ i = j, \\ 0, & \text{otherwise.} \end{cases}$$

where $r = 1, 2, \cdots, \sigma$.

Theorem 9.6 *The network* (9.28) *achieves output synchronization under the adaptive law* (9.29).

Proof For the network (9.30), the Lyapunov functional is chosen as follows:

$$V_3(t) = \sum_{i=1}^N e_i^T(t)Qe_i(t) + \sum_{r=1}^\sigma \sum_{i=1}^N \sum_{(i,j)\in\hat{E}} \frac{a_r}{4\beta^r_{ij}}(G^r_{ij}(t) - \hat{G}^r_{ij})^2, \tag{9.31}$$

where $\hat{G}^r_{ij} = \hat{G}^r_{ji}(i \neq j, i, j = 1, 2, \cdots, N)$ are nonnegative constants, and $\hat{G}^r_{ij} = 0 (i \neq j)$ if and only if $\dot{G}^r_{ij}(t)$=0.

Then, we get

$$\dot{V}_3(t) = 2\sum_{i=1}^{N} e_i^T(t)Q\Big[f(x_i(t)) - \frac{1}{N}\sum_{k=1}^{N} f(x_k(t)) + f(\bar{x}(t))$$
$$-f(\bar{x}(t)) + \sum_{r=1}^{\sigma}\sum_{j=1}^{N} a_r G_{ij}^r(t)\Gamma_r e_j(t)\Big]$$
$$+\frac{1}{2}\sum_{r=1}^{\sigma}\sum_{i=1}^{N}\sum_{(i,j)\in\hat{E}} a_r(G_{ij}^r(t) - \hat{G}_{ij}^r)(e_i(t)$$
$$-e_j(t))^T(Q\Gamma_r + \Gamma_r Q)(e_i(t) - e_j(t)). \tag{9.32}$$

Define $\hat{G}^r = (\hat{G}_{ij}^r)_{N\times N} \in \mathbb{R}^{N\times N}$, where $\hat{G}_{ii}^r = -\sum_{\substack{j=1\\ j\neq i}}^{N} \hat{G}_{ij}^r, i = 1, 2, \cdots, N$. Subsequently, one has

$$\sum_{i=1}^{N}\sum_{(i,j)\in\hat{E}} (G_{ij}^r(t) - \hat{G}_{ij}^r)(e_i(t) - e_j(t))^T(Q\Gamma_r + \Gamma_r Q)(e_i(t) - e_j(t))$$
$$= \sum_{i=1}^{N}\sum_{\substack{j=1\\ j\neq i}}^{N}(\Psi_{ij}^r(t) - \hat{G}_{ij}^r)(e_i(t) - e_j(t))^T(Q\Gamma_r + \Gamma_r Q)(e_i(t) - e_j(t))$$
$$= -2\sum_{i=1}^{N}\sum_{j=1}^{N}(\Psi_{ij}^r(t) - \hat{G}_{ij}^r)e_i^T(t)(Q\Gamma_r + \Gamma_r Q)e_j(t). \tag{9.33}$$

By use of (9.33), we know

$$\dot{V}_3(t) \leqslant 2\sum_{i=1}^{N} e_i^T(t)(Q\Xi - \eta I_n)e_i(t) + \sum_{r=1}^{\sigma}\sum_{i=1}^{N}\sum_{j=1}^{N} a_r\hat{G}_{ij}^r e_i^T(t)(Q\Gamma_r + \Gamma_r Q)e_j(t)$$
$$+\sum_{r=1}^{\sigma}\sum_{i=1}^{N}\sum_{j=1}^{N} a_r O_{ij}^r e_i^T(t)(Q\Gamma_r + \Gamma_r Q)e_j(t)$$
$$= e^T(t)\Big\{2I_N \otimes (Q\Xi - \eta I_n) + a_s\hat{G}^s \otimes (Q\Gamma_s + \Gamma_s Q) + \sum_{\substack{r=1\\ r\neq s}}^{\sigma} a_r\hat{G}^r \otimes (Q\Gamma_r$$
$$+\Gamma_r Q) + \sum_{r=1}^{\sigma} a_r O^r \otimes (Q\Gamma_r + \Gamma_r Q)\Big\}e(t)$$

$$\leqslant e^T(t)\Big\{2I_N \otimes (Q\Xi - \eta I_n) + a_s\hat{G}^s \otimes (Q\Gamma_s$$
$$+\Gamma_s Q) + \sum_{r=1}^{\sigma} a_r O^r \otimes (Q\Gamma_r + \Gamma_r Q)\Big\}e(t), \tag{9.34}$$

where $e(t) = (e_1^T(t), e_2^T(t), \cdots, e_N^T(t))^T$, $O^r = (O_{ij}^r)_{N\times N} \in \mathbb{R}^{N\times N}$.

Obviously, there exists a unitary matrix $P = (p_1, p_2, \cdots, p_N) \in \mathbb{R}^{N\times N}$ such that

$$P^T\hat{G}^s P = \lambda = \text{diag}(\lambda_1, \lambda_2, \cdots, \lambda_N) \in \mathbb{R}^{N\times N},$$

where $0 = \lambda_1 > \lambda_2 \geqslant \lambda_3 \geqslant \cdots \geqslant \lambda_N$. Let $z(t) = (z_1^T(t), z_2^T(t), \cdots, z_N^T(t))^T = (P^T \otimes I_n)e(t)$. Since $p_1 = \frac{1}{\sqrt{N}}(1, 1, \cdots, 1)^T$, one gets $z_1(t) = (p_1^T \otimes I_n)e(t) = 0$. Then, we can obtain from (9.34) that

$$\begin{aligned}
\dot{V}_3(t) &\leqslant e^T(t)\Big\{2I_N \otimes (Q\Xi - \eta I_n) + \sum_{r=1}^{\sigma} a_r O^r \otimes (Q\Gamma_r + \Gamma_r Q) \\
&\quad +a_s(P \otimes I_n)[\lambda \otimes (Q\Gamma_s + \Gamma_s Q)](P^T \otimes I_n)\Big\}e(t) \\
&= e^T(t)\Big\{2I_N \otimes (Q\Xi - \eta I_n) + \sum_{r=1}^{\sigma} a_r O^r \otimes (Q\Gamma_r \\
&\quad +\Gamma_r Q)\Big\}e(t) + a_s z^T(t)\big[\lambda \otimes (Q\Gamma_s + \Gamma_s Q)\big]z(t) \\
&\leqslant e^T(t)\Big\{2I_N \otimes (Q\Xi - \eta I_n) + \sum_{r=1}^{\sigma} a_r O^r \otimes (Q\Gamma_r \\
&\quad +\Gamma_r Q)\Big\}e(t) + a_s\lambda_2 z^T(t)\big[I_N \otimes (Q\Gamma_s + \Gamma_s Q)\big]z(t) \\
&= e^T(t)\Big\{2I_N \otimes (Q\Xi - \eta I_n) + \sum_{r=1}^{\sigma} a_r O^r \otimes (Q\Gamma_r \\
&\quad +\Gamma_r Q) + a_s\lambda_2\big[I_N \otimes (Q\Gamma_s + \Gamma_s Q)\big]\Big\}e(t).
\end{aligned} \tag{9.35}$$

Select sufficiently large $\hat{G}_{ij}^s$ such that

$$2\lambda_M\big(Q\Xi - \eta I_n\big) + \beta_1 + \beta_2 + \xi\lambda_M\big(\hat{C}^T\hat{C}\big) \leqslant 0, \tag{9.36}$$

where $\beta_1 = \lambda_M\big(\sum_{r=1}^{\sigma} a_r O^r \otimes (Q\Gamma_r + \Gamma_r Q)\big)$, $\beta_2 = a_s\lambda_2\lambda_m\big(Q\Gamma_s + \Gamma_s Q\big)$.

From (9.35) and (9.36), one gets

$$\dot{V}_3(t) \leqslant -\xi \sum_{i=1}^{N} e_i^T(t)\hat{C}^T\hat{C}e_i(t).$$

Therefore, $V_3(t)$ is nonincreasing, $e_i(t), i = 1, 2, \cdots, N$, and $G_{ij}^r(t), (i, j) \in \hat{E}$ are bounded. Namely,

$$\begin{aligned} \|e_i(t)\|_2 &\leqslant \varphi_i \in \mathbb{R} > 0, \\ |G_{ij}^r(t)| &\leqslant \chi_{ij}^r \in \mathbb{R} > 0 \end{aligned} \tag{9.37}$$

for any $t \in [0, +\infty)$.

Then, using a similar method as in Theorem 9.4, one gets

$$\lim_{t\to+\infty} \|\hat{y}_i^*(t)\|_2 = \lim_{t\to+\infty} \|\hat{C}e_i(t)\|_2 = 0,$$

where $i = 1, 2, \cdots, N$.

Consequently, we can deduce that the network (9.28) under the adaptive law (9.29) is output synchronized. □

9.3 $\mathcal{H}_\infty$ Output Synchronization and Pinning Control of Complex Networks with Multi-weights

9.3.1 Network Model

The multi-weighted complex network with external disturbances to be investigated in this section is given as follows:

$$\begin{aligned} \dot{x}_i(t) &= f(x_i(t)) + a_1 \sum_{j=1}^{N} G_{ij}^1 \Gamma_1 x_j(t) + a_2 \sum_{j=1}^{N} G_{ij}^2 \Gamma_2 x_j(t) \\ &\quad + \cdots + a_\sigma \sum_{j=1}^{N} G_{ij}^\sigma \Gamma_\sigma x_j(t) + \eta_i(t), \\ y_i(t) &= Cx_i(t), \end{aligned} \tag{9.38}$$

where $i = 1, 2, \cdots, N$, $x_i(t)$, $f(\cdot)$, a_r, Γ_r, $G^r = (G_{ij}^r)_{N\times N}$, C, $y_i(t)$ represent the same meanings as these in network (9.1); $\eta_i(t) \in \mathbb{R}^n$ is external disturbance, which is square integrable, i.e., $\int_0^{t_m} \eta_i^T(t)\eta_i(t)dt < +\infty$ for any $\mathbb{R} \ni t_m \geqslant 0$. In this section, the complex network (9.38) is also always required to be connected, and it has the same topology structure for different coupling forms.

9.3.2 $\mathcal{H}_\infty$ Output Synchronization

Letting $\bar{x}(t) = \frac{1}{N}\sum_{i=1}^{N} x_i(t)$, one has

$$\dot{\bar{x}}(t) = \frac{1}{N}\sum_{i=1}^{N} f(x_i(t)) + \frac{1}{N}\sum_{i=1}^{N} \eta_i(t).$$

Take $e_i(t) = x_i(t) - \bar{x}(t)$. Then, one gets

$$\begin{aligned}
\dot{e}_i(t) &= f(x_i(t)) - \frac{1}{N}\sum_{k=1}^{N} f(x_k(t)) + a_1\sum_{j=1}^{N} G_{ij}^1\Gamma_1 e_j(t) + a_2\sum_{j=1}^{N} G_{ij}^2\Gamma_2 e_j(t) \\
&\quad + \cdots + a_\sigma\sum_{j=1}^{N} G_{ij}^\sigma\Gamma_\sigma e_j(t) + \eta_i(t) - \frac{1}{N}\sum_{k=1}^{N}\eta_k(t), \\
y_i^*(t) &= Ce_i(t), \qquad (9.39)
\end{aligned}$$

where $y_i^*(t) = y_i(t) - C\bar{x}(t)$.

Obviously, there exist nonsingular matrices $M_1 \in \mathbb{R}^{p\times p}$ and $M_2 \in \mathbb{R}^{n\times n}$ such that

$$C = M_1\begin{pmatrix} I_b & 0 \\ 0 & 0 \end{pmatrix} M_2.$$

Then, we have

$$\begin{aligned}
\dot{e}_i(t) &= f(x_i(t)) - \frac{1}{N}\sum_{k=1}^{N} f(x_k(t)) + a_1\sum_{j=1}^{N} G_{ij}^1\Gamma_1 e_j(t) + a_2\sum_{j=1}^{N} G_{ij}^2\Gamma_2 e_j(t) \\
&\quad + \cdots + a_\sigma\sum_{j=1}^{N} G_{ij}^\sigma\Gamma_\sigma e_j(t) + \eta_i(t) - \frac{1}{N}\sum_{k=1}^{N}\eta_k(t), \\
\hat{y}_i^*(t) &= \hat{C}e_i(t), \qquad (9.40)
\end{aligned}$$

where $\hat{y}_i^*(t) = (\hat{y}_{i1}^*(t), \hat{y}_{i2}^*(t), \cdots, \hat{y}_{ib}^*(t))^T \in \mathbb{R}^b, \hat{C} = (I_b \quad 0)_{b\times n}M_2, i = 1, 2, \cdots, N.$

Definition 9.7 The network (9.38) is $\mathcal{H}_\infty$ output synchronized if

$$\sum_{i=1}^{N}\int_0^{t_m} (\hat{y}_i^*(t))^T\hat{y}_i^*(t)dt \leqslant V(0) + \mu^2\sum_{i=1}^{N}\int_0^{t_m}\eta_i^T(t)\eta_i(t)dt$$

for any $\mathbb{R} \ni t_m \geqslant 0$ and some nonnegative function $V(\cdot)$, where $\mathbb{R} \ni \mu > 0$.

Theorem 9.8 *If there exist two positive real numbers ξ and μ such that*

$$I_N \otimes \left[2(Q\Xi - \eta I_n) + \frac{Q^2}{\mu^2} + \hat{C}^T\hat{C}\right] + W_2 \leqslant 0, \tag{9.41}$$

where $W_2 = \sum_{r=1}^{\sigma} a_r G^r \otimes (Q\Gamma_r + \Gamma_r Q)$, then the network (9.38) is $\mathcal{H}_\infty$ output synchronized.

Proof Select the following Lyapunov functional for the network (9.40):

$$V_1(t) = \sum_{i=1}^{N} e_i^T(t) Q e_i(t).$$

Then, one has

$$\begin{aligned}
\dot{V}_1(t) = {} & 2\sum_{i=1}^{N} e_i^T(t) Q\Big[f(x_i(t)) - \frac{1}{N}\sum_{k=1}^{N} f(x_k(t)) + f(\bar{x}(t)) - f(\bar{x}(t)) - \frac{1}{N}\sum_{k=1}^{N} \eta_k(t) \\
& + a_1\sum_{j=1}^{N} G_{ij}^1 \Gamma_1 e_j(t) + a_2\sum_{j=1}^{N} G_{ij}^2 \Gamma_2 e_j(t) + \cdots + a_\sigma \sum_{j=1}^{N} G_{ij}^\sigma \Gamma_\sigma e_j(t) + \eta_i(t)\Big] \\
\leqslant {} & e^T(t)\Big[2I_N \otimes (Q\Xi - \eta I_n) + \sum_{r=1}^{\sigma} a_r G^r \otimes (Q\Gamma_r + \Gamma_r Q)\Big] e(t) \\
& + 2\sum_{i=1}^{N} e_i^T(t) Q \eta_i(t),
\end{aligned} \tag{9.42}$$

where $e(t) = (e_1^T(t), e_2^T(t), \cdots, e_N^T(t))^T$.

From (9.42), one gets

$$\begin{aligned}
& \dot{V}_1(t) + \sum_{i=1}^{N} (\hat{y}_i^*(t))^T \hat{y}_i^*(t) - \mu^2 \sum_{i=1}^{N} \eta_i^T(t)\eta_i(t) \\
\leqslant {} & e^T(t)\Big[2I_N \otimes (Q\Xi - \eta I_n) + \sum_{r=1}^{\sigma} a_r G^r \otimes (Q\Gamma_r + \Gamma_r Q)\Big] e(t) \\
& + 2\sum_{i=1}^{N} e_i^T(t) Q\eta_i(t) + \sum_{i=1}^{N} e_i^T(t)\hat{C}^T\hat{C} e_i(t) - \mu^2 \sum_{i=1}^{N} \eta_i^T(t)\eta_i(t) \\
\leqslant {} & e^T(t)\Big\{ I_N \otimes \Big[2(Q\Xi - \eta I_n) + \frac{Q^2}{\mu^2} + \hat{C}^T\hat{C}\Big]
\end{aligned}$$

$$+\sum_{r=1}^{\sigma} a_r G^r \otimes (Q\Gamma_r + \Gamma_r Q)\Big\}e(t). \tag{9.43}$$

From (9.41), we have

$$\sum_{i=1}^{N}(\hat{y}_i^*(t))^T \hat{y}_i^*(t) \leqslant -\dot{V}_1(t) + \mu^2 \sum_{i=1}^{N} \eta_i^T(t)\eta_i(t). \tag{9.44}$$

From (9.44), one obtains

$$\begin{aligned}&\sum_{i=1}^{N}\int_0^{t_m} (\hat{y}_i^*(t))^T \hat{y}_i^*(t)dt \\ &\leqslant V_1(0) - V_1(t_m) + \mu^2 \sum_{i=1}^{N}\int_0^{t_m} \eta_i^T(t)\eta_i(t)dt \\ &\leqslant V_1(0) + \mu^2 \sum_{i=1}^{N}\int_0^{t_m} \eta_i^T(t)\eta_i(t)dt\end{aligned}$$

for any $\mathbb{R} \ni t_m \geqslant 0$. Therefore, the complex network (9.38) is $\mathcal{H}_\infty$ output synchronized. □

9.3.3 Nodes-Based Pinning $\mathcal{H}_\infty$ Output Synchronization

Let the first l $(1 \leqslant l < N)$ nodes be pinned for realizing $\mathcal{H}_\infty$ output synchronization in complex network (9.38). Subsquently, the controlled network is given as follows:

$$\begin{aligned}\dot{x}_i(t) =& f(x_i(t)) + a_1 \sum_{j=1}^{N} G_{ij}^1 \Gamma_1 x_j(t) + a_2 \sum_{j=1}^{N} G_{ij}^2 \Gamma_2 x_j(t) + \cdots \\ &+ a_\sigma \sum_{j=1}^{N} G_{ij}^\sigma \Gamma_\sigma x_j(t) + u_i(t) + \eta_i(t), \; i = 1, 2, \cdots, l, \\ \dot{x}_i(t) =& f(x_i(t)) + a_1 \sum_{j=1}^{N} G_{ij}^1 \Gamma_1 x_j(t) + a_2 \sum_{j=1}^{N} G_{ij}^2 \Gamma_2 x_j(t) + \cdots \\ &+ a_\sigma \sum_{j=1}^{N} G_{ij}^\sigma \Gamma_\sigma x_j(t) + \eta_i(t), \; i = l+1, l+2, \cdots, N,\end{aligned}$$

$$y_i(t) = Cx_i(t), i = 1, 2, \cdots, N, \tag{9.45}$$

in which

$$\begin{aligned} u_i(t) &= -\sum_{r=1}^{\sigma} a_r h_i^r(t)\Gamma_r \left(x_i(t) - \frac{1}{N}\sum_{k=1}^{N} x_k(t) \right), \\ \dot{h}_i^r(t) &= \beta_i^r \left(x_i(t) - \frac{1}{N}\sum_{k=1}^{N} x_k(t) \right)^T \frac{Q\Gamma_r + \Gamma_r Q}{2} \\ &\quad \times \left(x_i(t) - \frac{1}{N}\sum_{k=1}^{N} x_k(t) \right), \end{aligned} \tag{9.46}$$

where $Q\Gamma_r + \Gamma_r Q > 0, \mathbb{R} \ni \beta_i^r > 0, \mathbb{R} \ni h_i^r(0) > 0, i = 1, 2, \cdots, l, \mathbb{R} \ni h_i^r(t) \equiv 0$ for $i = l+1, l+2, \cdots, N$.

Letting $\bar{x}(t) = \frac{1}{N}\sum_{i=1}^{N} x_i(t)$ and $e_i(t) = x_i(t) - \bar{x}(t)$, we can get

$$\begin{aligned} \dot{e}_i(t) &= f(x_i(t)) - \frac{1}{N}\sum_{k=1}^{N} f(x_k(t)) + \eta_i(t) - \frac{1}{N}\sum_{k=1}^{N} \eta_k(t) - \frac{1}{N}\sum_{k=1}^{l} u_k(t) \\ &\quad + \sum_{r=1}^{\sigma}\sum_{j=1}^{N} a_r G_{ij}^r \Gamma_r e_j(t) - \sum_{r=1}^{\sigma} a_r h_i^r(t)\Gamma_r e_i(t), \\ \hat{y}_i^*(t) &= \hat{C} e_i(t), i = 1, 2, \cdots, N, \\ \dot{h}_i^r(t) &= \beta_i^r e_i^T(t) \frac{Q\Gamma_r + \Gamma_r Q}{2} e_i(t), i = 1, 2, \cdots, l, \end{aligned} \tag{9.47}$$

where $\hat{C}$ and $\hat{y}_i^*(t)$ denote the same meanings as these in (9.40).

Theorem 9.9 *If there exist matrices* $H^r = \mathrm{diag}(\hat{h}_1^r, \hat{h}_2^r, \cdots, \hat{h}_l^r, 0, \cdots, 0) \in \mathbb{R}^{N\times N}, r = 1, 2, \cdots, \sigma$, *and a positive real number* μ *such that*

$$I_N \otimes W_3 + \sum_{r=1}^{\sigma} a_r (G^r - H^r) \otimes (Q\Gamma_r + \Gamma_r Q) \leqslant 0, \tag{9.48}$$

where $W_3 = 2(Q\Xi - \eta I_n) + \frac{Q^2}{\mu^2} + \hat{C}^T\hat{C}, \mathbb{R} \ni \hat{h}_i^r > 0, i = 1, 2, \cdots, l$, *then the network* (9.45) *is* $\mathcal{H}_\infty$ *output synchronized.*

Proof For the network (9.47), the Lyapunov functional is chosen as follows:

$$V_2(t) = \sum_{i=1}^{N} e_i^T(t) Q e_i(t) + \sum_{r=1}^{\sigma}\sum_{i=1}^{l} \frac{a_r}{\beta_i^r}(h_i^r(t) - \hat{h}_i^r)^2.$$

Then, we can acquire

$$\begin{aligned}
\dot{V}_2(t) = {} & 2\sum_{i=1}^{N} e_i^T(t) Q\Big[f(x_i(t)) - \frac{1}{N}\sum_{k=1}^{N} f(x_k(t)) + f(\bar{x}(t)) - f(\bar{x}(t)) + \eta_i(t) \\
& - \frac{1}{N}\sum_{k=1}^{N} \eta_k(t) + \sum_{r=1}^{\sigma}\sum_{j=1}^{N} a_r G_{ij}^r \Gamma_r e_j(t) - \frac{1}{N}\sum_{k=1}^{l} u_k(t) \\
& - \sum_{r=1}^{\sigma} a_r h_i^r(t) \Gamma_r e_i(t)\Big] + \sum_{r=1}^{\sigma}\sum_{i=1}^{N} a_r (h_i^r(t) - \hat{h}_i^r) e_i^T(t)(Q\Gamma_r + \Gamma_r Q) e_i(t) \\
\leqslant {} & e^T(t)\Big[2I_N \otimes (Q\Xi - \eta I_n) + \sum_{r=1}^{\sigma} a_r G^r \otimes (Q\Gamma_r + \Gamma_r Q) \\
& - \sum_{r=1}^{\sigma} a_r H^r \otimes (Q\Gamma_r + \Gamma_r Q)\Big] e(t) + 2\sum_{i=1}^{N} e_i^T(t) Q \eta_i(t),
\end{aligned} \tag{9.49}$$

where $e(t) = (e_1^T(t), e_2^T(t), \cdots, e_N^T(t))^T$, $\hat{h}_i^r = 0, i = l+1, l+2, \cdots, N$.

From (9.49), one has

$$\begin{aligned}
& \dot{V}_2(t) + \sum_{i=1}^{N} (\hat{y}_i^*(t))^T \hat{y}_i^*(t) - \mu^2 \sum_{i=1}^{N} \eta_i^T(t)\eta_i(t) \\
\leqslant {} & e^T(t)\Big[2I_N \otimes (Q\Xi - \eta I_n) + \sum_{r=1}^{\sigma} a_r G^r \otimes (Q\Gamma_r + \Gamma_r Q) \\
& - \sum_{r=1}^{\sigma} a_r H^r \otimes (Q\Gamma_r + \Gamma_r Q)\Big] e(t) + 2\sum_{i=1}^{N} e_i^T(t) Q \eta_i(t) \\
& + \sum_{i=1}^{N} e_i^T(t) \hat{C}^T \hat{C} e_i(t) - \mu^2 \sum_{i=1}^{N} \eta_i^T(t)\eta_i(t) \\
\leqslant {} & e^T(t)\Big\{ I_N \otimes \Big[2(Q\Xi - \eta I_n) + \frac{Q^2}{\mu^2} + \hat{C}^T \hat{C} \Big] \\
& + \sum_{r=1}^{\sigma} a_r (G^r - H^r) \otimes (Q\Gamma_r + \Gamma_r Q) \Big\} e(t).
\end{aligned} \tag{9.50}$$

From (9.50), we get

$$\sum_{i=1}^{N}\int_0^{t_m}(\hat{y}_i^*(t))^T\hat{y}_i^*(t)dt \leqslant V_2(0)+\mu^2\sum_{i=1}^{N}\int_0^{t_m}\eta_i^T(t)\eta_i(t)dt$$

for any $\mathbb{R}\ni t_m \geqslant 0$. Therefore, the network (9.45) is $\mathcal{H}_\infty$ output synchronized. □

9.3.4 Edges-Based Pinning $\mathcal{H}_\infty$ Output Synchronization

Suppose $E\subseteq\{1,2,\cdots,N\}\times\{1,2,\cdots,N\}$ represents the set of undirected edges in network (9.38), and network (9.38) is connected through $\hat{E}\subset E$. The network (9.38) with a portion of adaptive coupling weights is described by

$$\begin{aligned}\dot{x}_i(t) &= f(x_i(t))+a_1\sum_{j=1}^{N}G_{ij}^1(t)\Gamma_1x_j(t)+a_2\sum_{j=1}^{N}G_{ij}^2(t)\Gamma_2x_j(t)\\&\quad+\cdots+a_\sigma\sum_{j=1}^{N}G_{ij}^\sigma(t)\Gamma_\sigma x_j(t)+\eta_i(t),\\y_i(t) &= Cx_i(t),\ i=1,2,\cdots,N,\end{aligned}\tag{9.51}$$

where

$$\begin{aligned}&\dot{G}_{ij}^r(t)=\beta_{ij}^r(x_i(t)-x_j(t))^T(Q\Gamma_r+\Gamma_rQ)(x_i(t)-x_j(t)),\ \text{if}\ (i,j)\in\hat{E}\\&\dot{G}_{ij}^r(t)=0,\qquad\qquad \text{if}\ (i,j)\in E-\hat{E},\end{aligned}\tag{9.52}$$

in which $\mathbb{R}\ni\beta_{ij}^r=\beta_{ji}^r>0$, $Q\Gamma_r+\Gamma_rQ>0, r=1,2,\cdots,\sigma$.

Take $\bar{x}(t)=\frac{1}{N}\sum_{i=1}^{N}x_i(t)$ and $e_i(t)=x_i(t)-\bar{x}(t)$. Then, one has

$$\begin{aligned}\dot{e}_i(t) &= f(x_i(t))-\frac{1}{N}\sum_{k=1}^{N}f(x_k(t))+\eta_i(t)-\frac{1}{N}\sum_{k=1}^{N}\eta_k(t)\\&\quad+\sum_{r=1}^{\sigma}\sum_{j=1}^{N}a_rG_{ij}^r(t)\Gamma_re_j(t),\\\hat{y}_i^*(t) &= \hat{C}e_i(t),\ i=1,2,\cdots,N,\end{aligned}\tag{9.53}$$

where $\hat{C}$ and $\hat{y}_i^*(t)$ denote the same meanings as these in (9.40).

Theorem 9.10 *The network* (9.51) *is* $\mathcal{H}_\infty$ *output synchronized under the adaptive law* (9.52).

Proof For the network (9.53), the Lyapunov functional is chosen as follows:

$$V_3(t) = \sum_{i=1}^{N} e_i^T(t)Qe_i(t) + \sum_{r=1}^{\sigma}\sum_{i=1}^{N}\sum_{(i,j)\in\hat{E}} \frac{a_r}{4\beta_{ij}^r}(G_{ij}^r(t) - \hat{G}_{ij}^r)^2, \quad (9.54)$$

where $\hat{G}_{ij}^r = \hat{G}_{ji}^r (i \neq j, i, j = 1, 2, \cdots, N)$ are nonnegative constants, and $\hat{G}_{ij}^r = 0 (i \neq j)$ if and only if $\dot{G}_{ij}^r(t)$=0.

Then, we can derive

$$\begin{aligned}\dot{V}_3(t) &= 2\sum_{i=1}^{N} e_i^T(t)Q\Big[f(x_i(t)) - \frac{1}{N}\sum_{k=1}^{N} f(x_k(t)) + f(\bar{x}(t)) - f(\bar{x}(t)) + \eta_i(t) \\ &\quad -\frac{1}{N}\sum_{k=1}^{N}\eta_k(t) + \sum_{r=1}^{\sigma}\sum_{j=1}^{N} a_r G_{ij}^r(t)\Gamma_r e_j(t)\Big] \\ &\quad +\frac{1}{2}\sum_{r=1}^{\sigma}\sum_{i=1}^{N}\sum_{(i,j)\in\hat{E}} a_r(G_{ij}^r(t) - \hat{G}_{ij}^r)(e_i(t) - e_j(t))^T(Q\Gamma_r + \Gamma_r Q)(e_i(t) - e_j(t)) \\ &\leqslant e^T(t)\Big[2I_N \otimes (Q\Xi - \eta I_n) + \sum_{r=1}^{\sigma} a_r(\hat{G}^r + O^r) \otimes (Q\Gamma_r + \Gamma_r Q)\Big]e(t) \\ &\quad +2\sum_{i=1}^{N} e_i^T(t)Q\eta_i(t), \end{aligned} \quad (9.55)$$

where $e(t) = (e_1^T(t), e_2^T(t), \cdots, e_N^T(t))^T$, O^r and $\hat{G}^r$ denote the same meanings as these in Theorem 9.6.

Then, one gets

$$\begin{aligned}&\dot{V}_3(t) + \sum_{i=1}^{N}(\hat{y}_i^*(t))^T\hat{y}_i^*(t) - \mu^2\sum_{i=1}^{N}\eta_i^T(t)\eta_i(t) \\ &\leqslant e^T(t)\Big[2I_N \otimes (Q\Xi - \eta I_n) + \sum_{r=1}^{\sigma} a_r(\hat{G}^r + O^r) \otimes (Q\Gamma_r \\ &\quad +\Gamma_r Q)\Big]e(t) + 2\sum_{i=1}^{N} e_i^T(t)Q\eta_i(t) - \mu^2\sum_{i=1}^{N}\eta_i^T(t)\eta_i(t)\end{aligned}$$

$$
\begin{aligned}
&+\sum_{i=1}^{N} e_i^T(t)\hat{C}^T\hat{C}e_i(t) \\
&\leqslant e^T(t)\Big\{I_N \otimes \Big[2(Q\Xi - \eta I_n) + \frac{Q^2}{\mu^2} + \hat{C}^T\hat{C}\Big] \\
&\quad + \sum_{r=1}^{\sigma} a_r(\hat{G}^r + O^r) \otimes (Q\Gamma_r + \Gamma_r Q)\Big\}e(t) \\
&\leqslant e^T(t)\Big\{I_N \otimes \Big[2(Q\Xi - \eta I_n) + \frac{Q^2}{\mu^2} + \hat{C}^T\hat{C}\Big] \\
&\quad + a_s\hat{G}^s \otimes (Q\Gamma_s + \Gamma_s Q) + \sum_{r=1}^{\sigma} a_r O^r \otimes (Q\Gamma_r + \Gamma_r Q)\Big\}e(t),
\end{aligned}
$$

where $s \in \{1, 2, \cdots, \sigma\}$, $0 < \mu \in \mathbb{R}$.

Similar to the proof of Theorem 9.6, we can select sufficiently large $\hat{G}_{ij}^s$ such that

$$
W_4 + a_s\hat{G}^s \otimes (Q\Gamma_s + \Gamma_s Q) \leqslant 0, \tag{9.56}
$$

where $W_4 = I_N \otimes \Big[2(Q\Xi - \eta I_n) + \frac{Q^2}{\mu^2} + \hat{C}^T\hat{C}\Big] + \sum_{r=1}^{\sigma} a_r O^r \otimes (Q\Gamma_r + \Gamma_r Q)$.

From (9.56), one obtains

$$
\sum_{i=1}^{N}\int_0^{t_m}(\hat{y}_i^*(t))^T\hat{y}_i^*(t)dt \leqslant V_3(0) + \mu^2\sum_{i=1}^{N}\int_0^{t_m}\eta_i^T(t)\eta_i(t)dt
$$

for any $\mathbb{R} \ni t_m \geqslant 0$. Therefore, the network (9.51) is $\mathcal{H}_\infty$ output synchronized. □

9.4 Numerical Examples

Example 9.11 Consider a multi-weighted complex network described by:

$$
\begin{aligned}
\dot{x}_i(t) &= f(x_i(t)) + 2\sum_{j=1}^{6} G_{ij}^1\Gamma_1 x_j(t) + 3\sum_{j=1}^{6} G_{ij}^2\Gamma_2 x_j(t) + 4\sum_{j=1}^{6} G_{ij}^3\Gamma_3 x_j(t), \\
y_i(t) &= Cx_i(t),
\end{aligned}
$$

where $f_l(\xi) = 0.5\left(|\xi + 1| - |\xi - 1|\right)$, $l = 1, 2, 3$, $\Gamma_1 = \text{diag}(0.5, 0.7, 0.4)$, $\Gamma_2 = \text{diag}(0.6, 0.7, 0.5)$, $\Gamma_3 = \text{diag}(0.4, 0.5, 0.8)$, the matrices $C, G^1 = (G_{ij}^1)_{6\times 6}$, $G^2 = (G_{ij}^2)_{6\times 6}$ and $G^3 = (G_{ij}^3)_{6\times 6}$ are chosen as, respectively

$$C = \begin{pmatrix} 1 & 0 & 0 \\ 0 & 1 & 0 \end{pmatrix},$$

$$G^1 = \begin{pmatrix} -0.6 & 0.2 & 0 & 0.1 & 0.1 & 0.2 \\ 0.2 & -0.7 & 0.3 & 0.1 & 0.1 & 0 \\ 0 & 0.3 & -1.0 & 0 & 0.5 & 0.2 \\ 0.1 & 0.1 & 0 & -0.5 & 0.3 & 0 \\ 0.1 & 0.1 & 0.5 & 0.3 & -1.1 & 0.1 \\ 0.2 & 0 & 0.2 & 0 & 0.1 & -0.5 \end{pmatrix},$$

$$G^2 = \begin{pmatrix} -0.6 & 0.2 & 0 & 0.2 & 0.1 & 0.1 \\ 0.2 & -0.5 & 0.1 & 0.1 & 0.1 & 0 \\ 0 & 0.1 & -0.8 & 0 & 0.5 & 0.2 \\ 0.2 & 0.1 & 0 & -0.6 & 0.3 & 0 \\ 0.1 & 0.1 & 0.5 & 0.3 & -1.1 & 0.1 \\ 0.1 & 0 & 0.2 & 0 & 0.1 & -0.4 \end{pmatrix},$$

$$G^3 = \begin{pmatrix} -0.7 & 0.2 & 0 & 0.3 & 0.1 & 0.1 \\ 0.2 & -0.8 & 0.3 & 0.2 & 0.1 & 0 \\ 0 & 0.3 & -0.6 & 0 & 0.1 & 0.2 \\ 0.3 & 0.2 & 0 & -0.8 & 0.3 & 0 \\ 0.1 & 0.1 & 0.1 & 0.3 & -0.8 & 0.2 \\ 0.1 & 0 & 0.2 & 0 & 0.2 & -0.5 \end{pmatrix}.$$

Obviously, $f(\cdot)$ fulfills (A1) with $Q = I_3$, $\Xi = 2I_3$, $\eta = 1$. Choose nodes 1, 2, 3, 4 and 5 as pinning nodes.

Case 1: Using the MATLAB LMI Toolbox, we can get $\hat{h}_1^1 = 5.4831$, $\hat{h}_2^1 = 5.4046$, $\hat{h}_3^1 = 5.2364$, $\hat{h}_4^1 = 5.3414$, $\hat{h}_5^1 = 4.9497$, $\hat{h}_1^2 = 2.5457$, $\hat{h}_2^2 = 2.4868$, $\hat{h}_3^2 = 2.3943$, $\hat{h}_4^2 = 2.4656$, $\hat{h}_5^2 = 2.1620$, $\hat{h}_1^3 = 0.8133$, $\hat{h}_2^3 = 0.7642$, $\hat{h}_3^3 = 0.8179$, $\hat{h}_4^3 = 0.7790$, $\hat{h}_5^3 = 0.6256$, and $\xi = 0.4899$ satisfying (9.23). Based on Theorem 9.5, the network (9.20) under the controller (9.21) is output synchronized. Figs. 9.1 and 9.2 display the simulation results.

Case 2: Take $\eta_i(t) = (0.1i * \sin(t), 0.2i * \sin(t), 0.3i * \sin(t))^T$. By exploiting the MATLAB LMI Toolbox, we can obtain $\hat{h}_1^1 = 3.0889$, $\hat{h}_2^1 = 3.0269$, $\hat{h}_3^1 = 2.9171$, $\hat{h}_4^1 = 2.9779$, $\hat{h}_5^1 = 2.6972$, $\hat{h}_1^2 = 1.3075$, $\hat{h}_2^2 = 1.2603$, $\hat{h}_3^2 = 1.2097$, $\hat{h}_4^2 = 1.2529$, $\hat{h}_5^2 = 1.0143$, $\hat{h}_1^3 = 0.1659$, $\hat{h}_2^3 = 0.1252$, $\hat{h}_3^3 = 0.1906$, $\hat{h}_4^3 = 0.1629$, $\hat{h}_5^3 = 0.0353$, and $\mu = 1.7747$ satisfying (9.48). Based on Theorem 9.9, the network (9.45) under the controller (9.46) is $\mathcal{H}_\infty$ output synchronized. Figs. 9.3 and 9.4 display the simulation results.

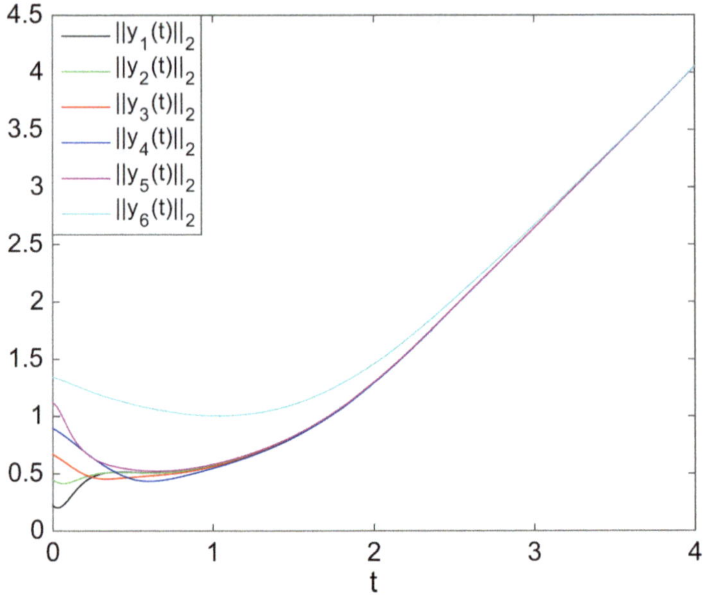

Fig. 9.1 $\|y_i(t)\|_2, i = 1, 2, \cdots, 6$

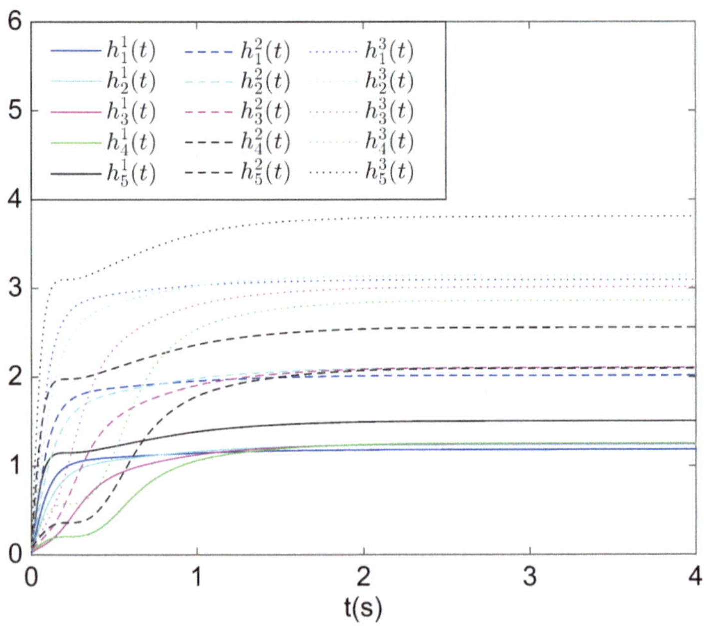

Fig. 9.2 $h_i^1(t), h_i^2(t), h_i^3(t), i = 1, 2, \cdots, 5$

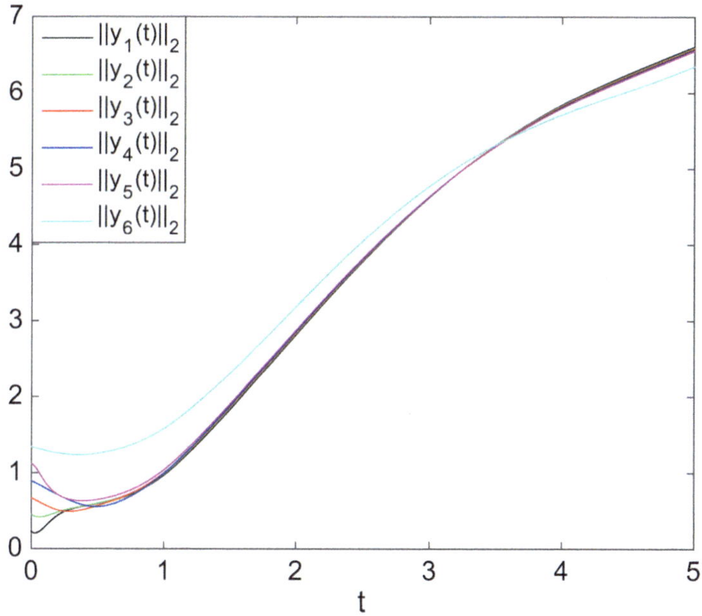

Fig. 9.3 $\|y_i(t)\|_2, i = 1, 2, \cdots, 6$

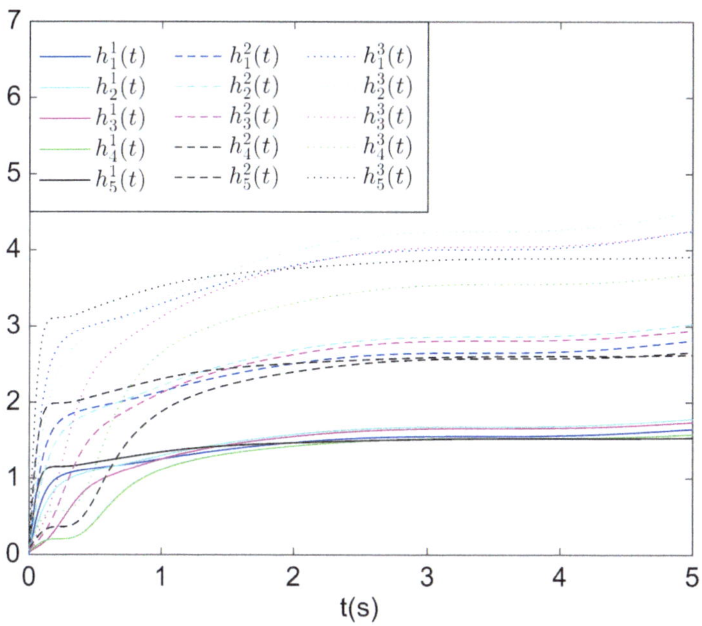

Fig. 9.4 $h_i^1(t), h_i^2(t), h_i^3(t), i = 1, 2, \cdots, 5$

Example 9.12 Consider a multi-weighted complex network described by:

$$\dot{x}_i(t)=f(x_i(t))+2\sum_{j=1}^{6}G_{ij}^1(t)\Gamma_1x_j(t)+3\sum_{j=1}^{6}G_{ij}^2(t)\Gamma_2x_j(t)+4\sum_{j=1}^{6}G_{ij}^3(t)\Gamma_3x_j(t),$$
$$y_i(t)=Cx_i(t),$$

where $f_1(\xi)=-0.3\tanh(\xi)$, $f_2(\xi)=-0.4\tanh(\xi)$, $f_3(\xi)=-0.5\tanh(\xi)$, $G^1(0)=(G_{ij}^1(0))_{6\times6}$, $G^2(0)=(G_{ij}^2(0))_{6\times6}$ and $G^3(0)=(G_{ij}^3(0))_{6\times6}$ are chosen as, respectively

$$\begin{pmatrix} -0.08 & 0.05 & 0.01 & 0.01 & 0.01 & 0 \\ 0.05 & -0.09 & 0.02 & 0 & 0.01 & 0.01 \\ 0.01 & 0.02 & -0.09 & 0.01 & 0.05 & 0 \\ 0.01 & 0 & 0.01 & -0.05 & 0.03 & 0 \\ 0.01 & 0.01 & 0.05 & 0.03 & -0.11 & 0.01 \\ 0 & 0.01 & 0 & 0 & 0.01 & -0.02 \end{pmatrix},$$

$$\begin{pmatrix} -0.06 & 0.02 & 0.01 & 0.01 & 0.02 & 0 \\ 0.02 & -0.05 & 0.01 & 0 & 0.01 & 0.01 \\ 0.01 & 0.01 & -0.08 & 0.01 & 0.05 & 0 \\ 0.01 & 0 & 0.01 & -0.05 & 0.03 & 0 \\ 0.02 & 0.01 & 0.05 & 0.03 & -0.12 & 0.01 \\ 0 & 0.01 & 0 & 0 & 0.01 & -0.02 \end{pmatrix},$$

$$\begin{pmatrix} -0.09 & 0.03 & 0.02 & 0.03 & 0.01 & 0 \\ 0.03 & -0.08 & 0.03 & 0 & 0.01 & 0.01 \\ 0.02 & 0.03 & -0.09 & 0.01 & 0.03 & 0 \\ 0.03 & 0 & 0.01 & -0.07 & 0.03 & 0 \\ 0.01 & 0.01 & 0.03 & 0.03 & -0.1 & 0.02 \\ 0 & 0.01 & 0 & 0 & 0.02 & -0.03 \end{pmatrix},$$

$\Gamma_r, r=1,2,3$ and C take the same values as in Example 9.11. Obviously, $f(\cdot)$ fulfills (A1) with $Q=I_3$, $\Xi=1.5I_3$, $\eta=0.5$. Select edges (1, 2), (2, 3), (3, 4), (4, 5), (5, 6) as pinned edges.

Case 1: According to Theorem 9.6, the network (9.28) under the adaptive law (9.29) is output synchronized. Figs. 9.5 and 9.6 display the simulation results.

Case 2: Take $\eta_i(t)=(0.05i*\sin(t), 0.1i*\sin(t), 0.15i*\sin(t))^T$. From Theorem 9.10, the network (9.51) under the adaptive law (9.52) is $\mathcal{H}_\infty$ output synchronized. Figs. 9.7 and 9.8 display the simulation results.

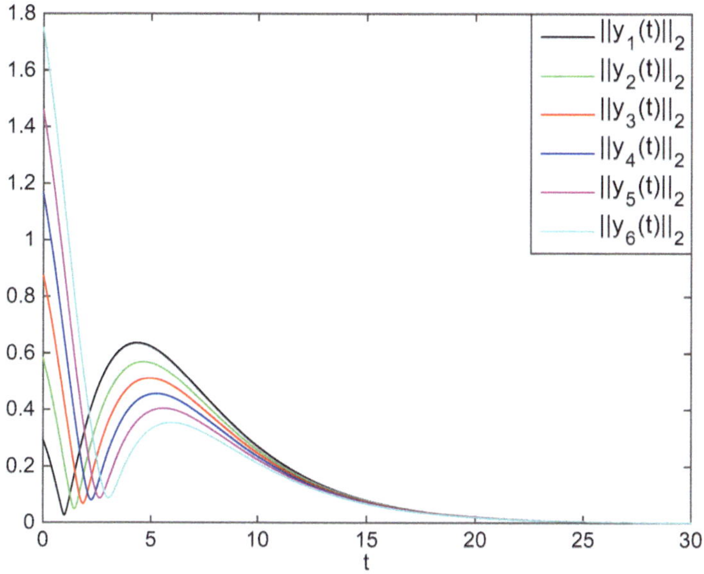

Fig. 9.5 $\|y_i(t)\|_2, i = 1, 2, \cdots, 6$

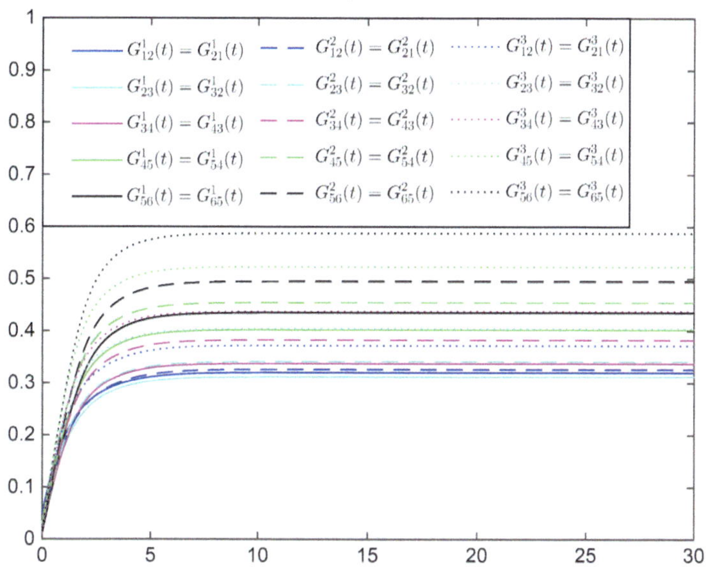

Fig. 9.6 Adaptive coupling weights

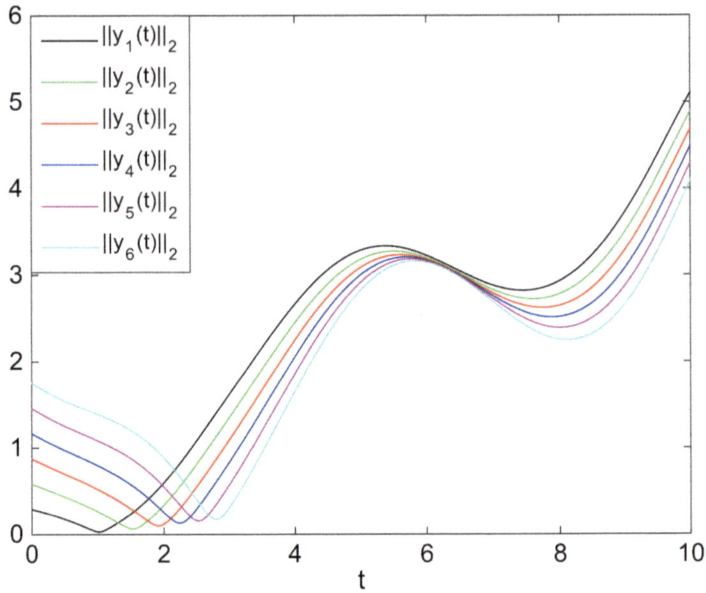

Fig. 9.7 $\|y_i(t)\|_2, i = 1, 2, \cdots, 6$

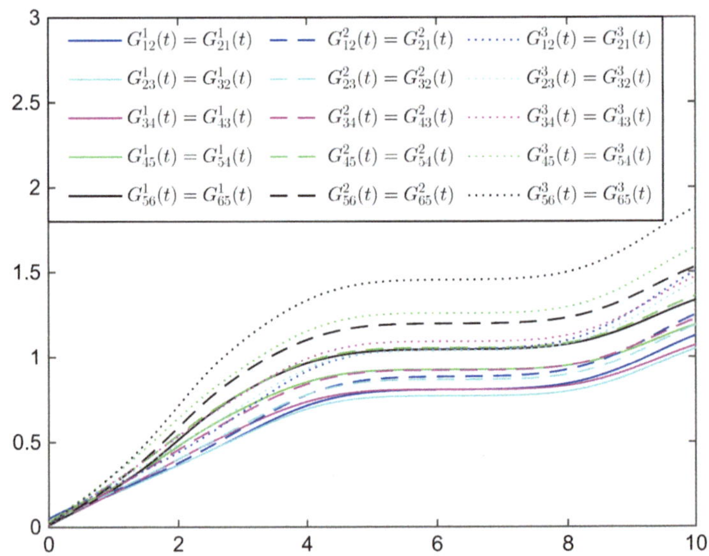

Fig. 9.8 Adaptive coupling weights

9.5 Conclusion

In this chapter, the multi-weighted CDNs with and without external disturbances have been discussed. With the help of the Lyapunov functionals and Barbalat's Lemma, we not only have analyzed the output synchronization of multi-weighted complex networks, but also have presented some nodes-based and edges-based pinning adaptive control strategies. On the other hand, a few similar results about the $\mathcal{H}_\infty$ output synchronization of multi-weighted complex networks with external disturbances have also been obtained. Finally, illustrative examples have been given to show the correctness of the derived results.

References

1. J.L. Wang, Z.C. Yang, H.N. Wu, Passivity analysis of complex dynamical networks with multiple time-varying delays. J. Eng. Math. **74**(1), 175–188 (2012)
2. S.H. Strogatz, Exploring complex networks. Nature **410**(6825), 268–276 (2001)
3. R. Albert, A. Barabási, Statistical mechanics of complex networks. Rev. Mod. Phys. **74**(1), 47–97 (2002)
4. H. Shen, X. Huang, J.P. Zhou, Z. Wang, Global exponential estimates for uncertain Markovian jump neural networks with reaction-diffusion terms. Nonlinear Dyn. **69**(1–2), 473–486 (2012)
5. J. Wang, J.H. Park, H. Shen, J. Wang, Delay-dependent robust dissipativity conditions for delayed neural networks with random uncertainties. Appl. Math. Comput. **221**, 710–719 (2013)
6. J. Zhou, J.A. Lu, J. Lü, Adaptive synchronization of an uncertain complex dynamical network. IEEE Trans. Autom. Control **51**(4), 652–656 (2006)
7. J. Zhou, J.A. Lu, J. Lü, Pinning adaptive synchronization of a general complex dynamical network. Automatica **44**(4), 996–1003 (2008)
8. J. Lü, G. Chen, A time-varying complex dynamical network model and its controlled synchronization criteria. IEEE Trans. Autom. Control **50**(6), 841–846 (2005)
9. W. Yu, G. Chen, J. Lü, On pinning synchronization of complex dynamical networks. Automatica **45**(2), 429–435 (2009)
10. W. Yu, G. Chen, M. Cao, Consensus in directed networks of agents with nonlinear dynamics. IEEE Trans. Autom. Control **56**(6), 1436–1441 (2011)
11. W. Yu, P. DeLellis, G. Chen, M.D. Bernardo, J. Kurths, Distributed adaptive control of synchronization in complex networks. IEEE Trans. Autom. Control **57**(8), 2153–2158 (2012)
12. M. Porfiri, M.D. Bernardo, Criteria for global pinning-controllability of complex networks. Automatica **44**(12), 3100–3106 (2008)
13. F. Sorrentino, M.D. Bernardo, F. Garofalo, G. Chen, Controllability of complex networks via pinning. Phys. Rev. E **75**(4), 046103 (2007)
14. B. Liu, D.J. Hill, Impulsive consensus for complex dynamical networks with nonidentical nodes and coupling time-delays. SIAM J. Control Optim. **49**(2), 315–338 (2011)
15. Z.G. Wu, P. Shi, H.Y. Su, J. Chu, Exponential synchronization of neural networks with discrete and distributed delays under time-varying sampling. IEEE Trans. Neural Netw. Learn. Syst. **23**(9), 1368–1376 (2012)
16. Z.G. Wu, P. Shi, H.Y. Su, J. Chu, Stochastic synchronization of Markovian jump neural networks with time-varying delay using sampled data. IEEE Trans. Cybern. **43**(6), 1796–1806 (2013)

J.-L. Wang et al., *Analysis and Control of Output Synchronization for Complex Dynamical Networks*, https://doi.org/10.1007/978-981-13-1352-3

17. Z.G. Wu, P. Shi, H.Y. Su, J. Chu, Sampled-data exponential synchronization of complex dynamical networks with time-varying coupling delay. IEEE Trans. Neural Netw. Learn. Syst. **24**(8), 1177–1187 (2013)
18. Y.W. Wang, H.O. Wang, J.W. Xiao, Z.H. Guan, Synchronization of complex dynamical networks under recoverable attacks. Automatica **46**(1), 197–203 (2010)
19. J.L. Wang, Z. Yang, T. Huang, M. Xiao, Local and global exponential synchronization of complex delayed dynamical networks with general topology. Discrete Contin. Dyn. Syst. Ser. B **16**(1), 393–408 (2012)
20. J.L. Wang, Z. Yang, T. Huang, M. Xiao, Synchronization criteria in complex dynamical networks with nonsymmetric coupling and multiple time-varying delays. Appl. Anal. **91**(5), 923–935 (2012)
21. T. Huang, G. Chen, J. Kurths, Synchronization of chaotic systems with time-varying coupling delays. Discrete Contin. Dyn. Syst. Ser. B **16**(4), 1071–1082 (2011)
22. G.P. Jiang, W.K. Tang, G. Chen, A state-observer-based approach for synchronization in complex dynamical networks. IEEE Trans. Circuits Syst. I: Reg. Pap. **53**(12), 2739–2745 (2006)
23. M.Y. Chen, Synchronization in complex dynamical networks with random sensor delay. IEEE Trans. Circuits Syst. Express Briefs **57**(1), 46–50 (2010)
24. J.L. Wang, H.N. Wu, Local and global exponential output synchronization of complex delayed dynamical networks. Nonlinear Dyn. **67**(1), 497–504 (2012)
25. J.L. Wang, H.N. Wu, Synchronization criteria for impulsive complex dynamical networks with time-varying delay. Nonlinear Dyn. **70**(1), 13–24 (2012)
26. Y.H. Zhao, J.L. Wang, Exponential synchronization of impulsive complex networks with output coupling. Int. J. Autom. Comput. **10**(4), 350–359 (2013)
27. T. Liu, D.J. Hill, J. Zhao, Output synchronization of dynamical networks with incrementally-dissipative nodes and switching topology. IEEE Trans. Circuits Syst. I: Reg. Pap. **62**(9), 2312–2323 (2015)
28. W.Y. Duan, X.R. Fu, Finite-time output synchronization for complex dynamical networks with time-varying delays, in *Proceedings of the 35th Chinese Control Conference*, Chengdu, 2016, pp. 1169–1174
29. J.L. Wang, H.N. Wu, T. Huang, S.Y. Ren, J. Wu, Passivity and output synchronization of complex dynamical networks with fixed and adaptive coupling strength. IEEE Trans. Neural Netw. Learn. Syst. **29**(2), 364–376 (2018)
30. J.L. Wang, H.N. Wu, Adaptive output synchronization of complex delayed dynamical networks with output coupling. Neurocomputing **142**, 174–181 (2014)
31. Q. Xia, G.P. Jiang, Output impulsive synchronization in complex dynamical networks with coupled delays, in *Proceedings of the 31st Chinese Control Conference*, Hefei, 2012, pp. 128–133
32. L.J. Zhu, Z.Y. Chen, R.H. Middleton, A general framework for robust output synchronization of heterogeneous nonlinear networked systems. IEEE Trans. Autom. Control **61**(8), 2092–2107 (2016)
33. H. Yu, P.J. Antsaklis, Output synchronization of networked passive systems with event-driven communication. IEEE Trans. Autom. Control **59**(3), 750–756 (2014)
34. Y.Q. Wu, R.Q. Lu, P. Shi, H.Y. Su, Z.G. Wu, Adaptive output synchronization of heterogeneous network with an uncertain leader. Automatica **76**, 183–192 (2017)
35. Y.Y. Liu, J. Zhao, Generalized output synchronization of dynamical networks using output quasi-passivity. IEEE Trans. Circuits Syst. I: Reg. Pap. **59**(6), 1290–1298 (2012)
36. L.Z. Gan, Output synchronization of a general complex dynamical network with time-varying delays, in *Proceedings of the 29th Chinese Control Conference*, Beijing, 2010, pp. 4696–4700
37. T. Huang, C. Li, S. Duan, J.A. Starzyk, Robust exponential stability of uncertain delayed neural networks with stochastic perturbation and impulse effects. IEEE Trans. Neural Netw. Learn. Syst. **23**(6), 866–875 (2012)

38. T. Huang, C. Li, W. Yu, G. Chen, Synchronization of delayed chaotic systems with parameter mismatches by using intermittent linear state feedback. Nonlinearity **22**(3), 569–584 (2009)
39. X.L. An, L. Zhang, J.G. Zhang, Research on urban public traffic network with multi-weights based on single bus transfer junction. Physica A **436**, 748–755 (2015)
40. X.L. An, L. Zhang, Y.Z. Li, J.G. Zhang, Synchronization analysis of complex networks with multi-weights and its application in public traffic network. Physica A **412**, 149–156 (2014)
41. Y.P. Zhao, P. He, H.S. Nik, J. Ren, Robust adaptive synchronization of uncertain complex networks with multiple time-varying coupled delays. Complexity **20**(6), 62–73 (2015)
42. J.C. Willems, Dissipative dynamical systems part I: general theory. Arch. Ration. Mech. Anal. **45**(5), 321–351 (1972)
43. W.L. Lu, T.P. Chen, New approach to synchronization analysis of linearly coupled ordinary differential systems. Physica D **213**(2), 214–230 (2006)
44. W.L. Guo, F. Austin, S.H. Chen, Global synchronization of nonlinearly coupled complex networks with non-delayed and delayed coupling. Commun. Nonlinear Sci. Numer. Simul. **15**(6), 1631–1639 (2010)
45. E.N. Lorenz, Deterministic nonperiodic flow. J. Atmos. Sci. **20**(2), 130–141 (1963)
46. G. Chen, T. Ueta, Yet another chaotic attractor. Int. J. Bifurcation Chaos **9**(07), 1465–1466 (1999)
47. J. Lü, G. Chen, A new chaotic attractor coined. Int. J. Bifurcation Chaos **12**(03), 659–661 (2002)
48. R.A. Brualdi, H.J. Ryser, *Combinatorial Matrix Theory*, vol. 39 (Cambridge University Press, Cambridge/New York, 1991)
49. A. Berman, R.J. Plemmons, *Nonnegative Matrices in Mathematical Sciences* (Academic, New York, 1979)
50. J.J.E. Slotine, W.P. Li, *Applied Nonlinear Control* (Prentice-Hall, Upper Saddle River, 1991)
51. J.L. Wang, H.N. Wu, T. Huang, S.Y. Ren, Passivity and synchronization of linearly coupled reaction-diffusion neural networks with adaptive coupling. IEEE Trans. Cybern. **45**(9), 1942–1952 (2015)
52. J. Chen, X. Chen, *Special Matrices* (Tsinghua University Press, Beijing, 2001)
53. W. Yu, G. Chen, J. Lü, J. Kurths, Synchronization via pinning control on general complex networks. SIAM J. Control. Optim. **51**(2) 1395–1416 (2013)
54. W. Yu, G. Chen, M. Cao, J. Kurths, Second-order consensus for multiagent systems with directed topologies and nonlinear dynamics. IEEE Trans. Syst. Man Cybern. B: Cybern. **40**(3), 881–891 (2010)
55. J.L. Wang, H.N. Wu, T. Huang, S.Y. Ren, J. Wu, Pinning control for synchronization of coupled reaction-diffusion neural networks with directed topologies. IEEE Trans. Syst. Man Cybern.: Syst. **46**(8), 1109–1120 (2016)
56. W. Yu, J. Cao, J. Lü, Global synchronization of linearly hybrid coupled networks with time-varying delay. SIAM J. Appl. Dyn. Syst. **7**(1), 108–133 (2008)
57. L.Y. Xiang, Z.Q. Chen, Z.X. Liu, F. Chen, Z.Z. Yuan, Pinning control of complex dynamical networks with heterogeneous delays. Comput. Math. Appl. **56**(5), 1423–1433 (2008)
58. L.Y. Xiang, Z.X. Liu, Z.Q. Chen, Z.Z. Yuan, Pinning weighted complex networks with heterogeneous delays by a small number of feedback controllers. Sci. China Ser. F Inf. Sci. **51**(5), 511–523 (2008)
59. W.L. Lu, F.M. Atay, J. Jost, Synchronization of discrete-time dynamical networks with time-varying couplings. SIAM J. Math. Anal. **39**(4), 1231–1259 (2008)
60. J. Yao, Z.H. Guan, D.J. Hill, Passivity-based control and synchronization of general complex dynamical networks. Automatica **45**(9), 2107–2113 (2009)
61. J. Yao, H.O. Wang, Z.H. Guan, W. Xu, Passive stability and synchronization of complex spatio-temporal switching networks with time delays. Automatica **45**(7), 1721–1728 (2009)
62. Z. Li, L.C. Jiao, J.J. Lee, Robust adaptive global synchronization of complex dynamical networks by adjusting time-varying coupling strength. Physica A **387**(5), 1369–1380 (2008)
63. S.J. Long, D.Y. Xu, Delay-dependent stability analysis for impulsive neural networks with time varying delays. Neurocomputing **71**(7–9), 1705–1713 (2008)

64. J. Zhou, L. Xiang, Z.R. Liu, K.A. Dawson, J.O. Indekeu, H.E. Stanley, C. Tsallis, Synchronization in complex delayed dynamical networks with impulsive effects. Physica A **384**(2), 684–692 (2007)
65. K. Li, C.H. Lai, Adaptive impulsive synchronization of uncertain complex dynamical networks. Phys. Lett. A **372**(10), 1601–1606 (2008)
66. S.M. Cai, J.Z. Zhou, L. Xiang, Z.R. Liu, Robust impulsive synchronization of complex delayed dynamical networks. Phys. Lett. A **372**(30), 4990–4995 (2008)
67. M. Chavez, D.U. Hwang, A. Amann, H.G.E. Hentschel, S. Boccaletti, Synchronization is enhanced in weighted complex networks. Phys. Rev. Lett. **94**(21), 218701 (2005)
68. A.E. Hramov, A.E. Khramova, A.A. Koronovshii, S. Boccaletti, Synchronization in networks of slightly nonidentical elements. Int. J. Bifurcation Chaos **18**(03), 845–850 (2008)
69. J.C. Willems, Dissipative dynamical systems part II: linear systems with quadratic supply rates. Arch. Ration. Mech. Anal. **45**(5), 352–393 (1972)
70. V. Belevitch, *Classical Network Synthesis* (Van Nostrand, New York, 1968)
71. D.J. Hill, P.J. Moylan, Stability results for nonlinear feedback systems. Automatica **13**(4), 377–382 (1997)
72. G.J. Santosuosso, Passivity of nonlinear systems with input-output feedthrough. Automatica **33**(4), 693–697 (1977)
73. L.O. Chua, Passivity and complexity. IEEE Trans. Circuits Syst. I: Fundam. Theory Appl. **46**(1), 71–82 (1999)
74. L. Xie, M. Fu, H. Li, Passivity analysis and passification for uncertain signal processing systems. IEEE Trans. Signal Process. **46**(9), 2394–2403 (1998)
75. W. Yu, Passive equivalence of chaos in lorenz system. IEEE Trans. Circuits Syst. I: Fundam. Theory Appl. **46**(7), 876–878 (1999)
76. C.W. Wu, Synchronization in arrays of coupled nonlinear systems: passivity, circle criterion, and observer design. IEEE Trans. Circuits Syst. I: Fundam. Theory Appl. **48**(10), 1257–1261 (2001)
77. G. Calcev, R. Gorez, M.D. Neyer, Passivity approach to fuzzy control systems. Automatica **34**(3), 339–344 (1998)
78. H.B. Zeng, Y. He, M. Wu, H.Q. Xiao, Improved conditions for passivity of neural networks with a time-varying delay. IEEE Trans. Cybern. **44**(6), 785–792 (2017)
79. Z.G. Wu, P. Shi, H. Su, J. Chu, Passivity analysis for discrete-time stochastic Markovian jump neural networks with mixed time delays. IEEE Trans. Neural Netw. **22**(10), 1566–1575 (2011)
80. H. Li, H. Gao, P. Shi, New passivity analysis for neural networks with discrete and distributed delays. IEEE Trans. Neural Netw. **21**(11), 1842–1847 (2010)
81. J.L. Wang, H.N. Wu, Z.C. Yang, Passivity analysis of impulsive complex networks. Int. J. Autom. Comput. **8**(4), 484–489 (2011)
82. J.L. Wang, H.N. Wu, L. Guo, Passivity and stability analysis of reaction-diffusion neural networks with dirichlet boundary conditions. IEEE Trans. Neural Netw. **22**(12), 2105–2116 (2011)
83. J.L. Wang, H.N. Wu, Robust stability and robust passivity of parabolic complex networks with parametric uncertainties and time-varying delays. Neurocomputing **87**, 26–32 (2012)
84. J.L. Wang, H.N. Wu, Passivity of delayed reaction-diffusion networks with application to a food web model. Appl. Math. Comput. **219**(24), 11311–11326 (2013)
85. D. Wu, S. Zhu, X. Luo, L. Wu, Effects of adaptive coupling on stochastic resonance of small-world networks. Phys. Rev. E **84**(2), 021102–1–021102–6 (2011)
86. H. Li, P. Shi, D. Yao, L. Wu, Observer-based adaptive sliding mode control for nonlinear Markovian jump systems. Automatica **64**, 133–142 (2016)
87. H. Li, L. Wang, H. Du, A. Boulkroune, Adaptive fuzzy backstepping tracking control for strict-feedback systems with input delay. IEEE Trans. Fuzzy Syst. **25**(3), 642–652 (2017)
88. J.L. Wang, H.N. Wu, L. Guo, Novel adaptive strategies for synchronization of linearly coupled neural networks with reaction-diffusion terms. IEEE Trans. Neural Netw. Learn. Syst. **25**(2), 429–440 (2014)

89. J.L. Wang, H.N. Wu, Synchronization and adaptive control of an array of linearly coupled reaction-diffusion neural networks with hybrid coupling. IEEE Trans. Cybern. **44**(8), 1350–1361 (2014)
90. J. Zhao, D.J. Hill, T. Liu, Synchronization of complex dynamical networks with switching topology: a switched system point of view. Automatica **45**(11), 2502–2511 (2009)
91. X. Liu, H. Su, M.Z.Q. Chen, A switching approach to designing finite-time synchronization controllers of coupled neural networks. IEEE Trans. Neural Netw. Learn. Syst. **27**(2), 471–482 (2016)
92. X. Yang, J. Cao, Z. Yang, Synchronization of coupled reaction-diffusion neural networks with time-varying delays via pinning-impulsive controller. SIAM J. Control Optim. **51**(5), 3486–3510 (2013)
93. E. Steur, H. Nijmeijer, Synchronization in networks of diffusively time-delay coupled (semi-)passive systems. IEEE Trans. Circuits Syst. I: Reg. Pap. **58**(6), 1358–1371 (2011)
94. Y.C. Liu, N. Chopra, Controlled synchronization of heterogeneous robotic manipulators in the task space. IEEE Trans. Robot. **28**(1), 268–275 (2012)
95. E. Steur, I. Tyukin, H. Nijmeijer, Semi-passivity and synchronization of diffusively coupled neuronal oscillators. Physica D **238**(21), 2119–2128 (2009)
96. D. Hill, P. Moylan, The stability of nonlinear dissipative systems. IEEE Trans. Autom. Control **21**(5), 708–711 (1976)
97. H. Li, P. Shi, D. Yao, Adaptive sliding-mode control of Markov jump nonlinear systems with actuator faults. IEEE Trans. Autom. Control **62**(4), 1933–1939 (2017)
98. N. Chopra, Output synchronization on strongly connected graphs. IEEE Trans. Autom. Control **57**(11), 2896–2901 (2012)
99. S. Kawamura, M. Svinin, *Advances in Robot Control* (Springer, Berlin, 2006)
100. P. Shi, F. Li, L. Wu, C.C. Lim, Neural network-based passive filtering for delayed neutral-type semi-Markovian jump systems. IEEE Trans. Neural Netw. Learn. Syst. **28**(9), 2101–2114 (2017)
101. F. Li, L. Wu, P. Shi, C.C. Lim, State estimation and sliding mode control for semi-Markovian jump systems with mismatched uncertainties. Automatica **51**, 385–393 (2015)
102. F. Li, P. Shi, L. Wu, X. Zhang, Fuzzy-model-based $\mathcal{D}$-stability and nonfragile control for discrete-time descriptor systems with multiple delays. IEEE Trans. Fuzzy Syst. **22**(4), 1019–1025 (2014)
103. Q. Song, J. Cao, On pinning synchronization of directed and undirected complex dynamical networks. IEEE Trans. Circuits Syst. I: Reg. Pap. **57**(3), 672–680 (2010)
104. B. Liu, W. Lu, T. Chen, Synchronization in complex networks with stochastically switching coupling structures. IEEE Trans. Autom. Control. **57**(3), 754–760 (2012)
105. N. Chopra, M.W. Spong, Passivity-based control of multi-agent systems, in *Advances in Robot Control: From Everyday Physics to Human-Like Movements* (Springer, Berlin, 2006)
106. H.F. Grip, T. Yang, A. Saberi, A.A. Stoorvogel, Output synchronization for heterogeneous networks of non-introspective agent. Automatica **48**(10), 2444–2453 (2012)
107. C. Zhou, J. Kurths, Dynamical weights and enhanced synchronization in adaptive complex networks. Phys. Rev. Lett. **96**(16), 164–102 (2006)
108. P.D. Lellis, M.D. Bernardo, F. Garofalo, Synchronization of complex networks through local adaptive coupling. Chaos **18**(3), 037–110 (2008)
109. P.D. Lellis, M.D. Bernardo, F. Garofalo, Novel decentralized adaptive strategies for the synchronization of complex networks. Automatica **45**(3), 1312–1318 (2009)
110. P.D. Lellis, M.D. Bernardo, F. Garofalo, Adaptive pinning control of networks of circuits and systems in Lur'e form. IEEE Trans. Circuits Syst. I: Reg. Pap. **60**(11), 3033–3042 (2013)
111. J.L. Wang, H.N. Wu, T.W. Huang, S.Y. Ren, Pinning control strategies for synchronization of linearly coupled neural networks with reaction-diffusion terms. IEEE Trans. Neural Netw. Learn. Syst. **27**(4), 749–761 (2016)
112. P. DeLellis, M.D. Bernardo, G. Russo, On quad, lipschitz, and contracting vector fields for consensus and synchronization of networks. IEEE Trans. Circuits Syst. I: Reg. Pap. **58**(3), 576–583 (2011)

113. M. Bürger, C.D. Persis, Further result about dynamic coupling for nonlinear output agreement, in *Proceedings of the 53rd IEEE Conference on Decision and Control*, Los Angeles (2015), pp. 1353–1358
114. P. Wieland, R. Sepulchre, F. Allgöwer, An internal model principle is necessary and sufficient for linear output synchronization. Automatica **47**(5), 1068–1074 (2011)
115. H. Kim, H. Shim, J.H. Seo, Output consensus of heterogeneous uncertain linear multi-agent systems. IEEE Trans. Autom. Control **56**(1), 200–206 (2011)
116. P. Wieland, J. Wu, F. Allgöwer, On synchronous steady states and internal models of diffusively coupled systems. IEEE Trans. Autom. Control **58**(10), 2591–2602 (2013)
117. A. Isidori, L. Marconi, G. Casadei, Robust output synchronization of a network of heterogeneous nonlinear agents via nonlinear regulation theory. IEEE Trans. Autom. Control **59**(10), 2680–2691 (2014)
118. H. Li, J. Wang, H.K. Lam, Q. Zhou, H. Du, Adaptive sliding mode control for interval type-2 fuzzy systems. IEEE Trans. Syst. Man. Cybern. Syst. **46**(12), 1654–1663 (2016)
119. Q. Zhou, P. Shi, S. Xu, H. Li, Observer-based adaptive neural network control for nonlinear stochastic systems with time delay. IEEE Trans. Neural Netw. Learn. Syst. **24**(1), 71–80 (2013)
120. Q. Zhou, P. Shi, Y. Tian, M. Wang, Approximation-based adaptive tracking control for mimo nonlinear systems with input saturation. IEEE Trans. Cybern. **45**(10), 2119–2128 (2015)
121. B. Yang, J. Wang, J. Wang, Stability analysis of delayed neural networks via a new integral inequality. Neural Netw. **88**, 49–57 (2017)
122. J. Wang, L. Tian, Global lagrange stability for inertial neural networks with mixed time-varying delays. Neurocomputing **235**, 140–146 (2017)
123. W.W. Che, W. Guan, Y.L. Wang, Adaptive regulation synchronization for a class of delayed Cohen–Grossberg neural networks. Nonlinear Dyn. **74**(4), 929–942 (2013)
124. Q. Chen, L. Liu, A. Wu, Mean-square global exponential stability in lagrange sense for delayed recurrent neural networks with Markovian switching. Neurocomputing **226**, 58–65 (2017)
125. Z.G. Wu, P. Shi, H. Su, J. Chu, Sampled-data synchronization of chaotic Lur'e systems with time delays. IEEE Trans. Neural Netw. Learn. Syst. **24**(3), 410–421 (2013)
126. Z. Cai, L. Huang, M. Zhu, D. Wang, Finite-time stabilization control of memristor-based neural networks. Nonlinear Anal. Hybrid Syst. **20**, 37–54 (2016)
127. L. Wei, W.H. Chen, G. Huang, Globally exponential stabilization of neural networks with mixed time delays via impulsive control. Appl. Math. Comput. **260**, 10–26 (2015)
128. J. Feng, S. Wang, Z. Wang, Stochastic synchronization in an array of neural networks with hybrid nonlinear coupling. Neurocomputing **74**(18), 3808–3815 (2011)
129. H. Wang, Q. Song, Synchronization for an array of coupled stochastic discrete-time neural networks with mixed delays. Neurocomputing **74**(10), 1572–1584 (2011)
130. Y. Tang, J.A. Fang, Adaptive synchronization in an array of chaotic neural networks with mixed delays and jumping stochastically hybrid coupling. Commun. Nonlinear Sci. Numer. Simul. **14**(9), 3615–3628 (2009)
131. Y. Tang, J.A. Fang, Robust synchronization in an array of fuzzy delayed cellular neural networks with stochastically hybrid coupling. Neurocomputing **72**(13), 3253–3262 (2009)
132. Y. Tang, Z. Wang, J. Fang, Controller design for synchronization of an array of delayed neural networks using a controllable probabilistic PSO. Inf. Sci. **181**(20), 4715–4732 (2011)
133. J. Cao, P. Li, W. Wang, Global synchronization in arrays of delayed neural networks with constant and delayed coupling. Phys. Lett. A **353**(4), 318–325 (2006)
134. W. Wang, J. Cao, Synchronization in an array of linearly coupled networks with time-varying delay. Physica A **366**, 197–211 (2006)
135. W. Yu, J. Cao, W. Lu, Synchronization control of switched linearly coupled neural networks with delay. Neurocomputing **73**(4), 858–866 (2010)
136. J. Lu, D.W.C. Ho, M. Liu, Globally exponential synchronization in an array of asymmetric coupled neural networks. Phys. Lett. A **369**(5-6), 444–451 (2007)

137. K. Yuan, Robust synchronization in arrays of coupled networks with delay and mixed coupling. Neurocomputing **72**(4-6), 1026–1031 (2009)
138. X. Yang, J. Cao, J. Lu, Synchronization of Markovian coupled neural networks with nonidentical node-delays and random coupling strengths. IEEE Trans. Neural Netw. Learn. Syst. **23**(1), 60–71 (2012)
139. H. Zhang, J. Wang, Z. Wang, H. Liang, Mode-dependent stochastic synchronization for Markovian coupled neural networks with time-varying mode-delays. IEEE Trans. Neural Netw. Learn. Syst. **26**(11), 2621–2634 (2015)
140. H. Chen, P. Shi, C.C. Lim, Exponential synchronization for Markovian stochastic coupled neural networks of neutral-type via adaptive feedback control. IEEE Trans. Neural Netw. Learn. Syst. **28**(7), 1618–1632 (2017)
141. J. Wang, H. Zhang, Z. Wang, Z. Liu, Sampled-data synchronization of Markovian coupled neural networks with mode delays based on mode-dependent LKF. IEEE Trans. Neural Netw. Learn. Syst. **28**(11), 2626–2637 (2017)
142. J. Wang, L. Su, H. Shen, Z.G. Wu, J.H. Park, Mixed H_∞/passive sampled-data synchronization control of complex dynamical networks with distributed coupling delay. J. Frankl. Inst. **354**(3), 1302–1320 (2017)
143. H. Shen, J.H. Park, Z.G. Wu, Z. Zhang, Finite-time H_∞ synchronization for complex networks with semi-Markov jump topology. Commun. Nonlinear Sci. Numer. Simul. **24**(1–3), 40–51 (2015)
144. L. Liu, W.H. Chen, X. Lu, Aperiodically intermittent synchronization for a class of reaction-diffusion neural networks. Neurocomputing **222**, 105–115 (2017)
145. W.H. Chen, Z. Jiang, X. Lu, S. Luo, H_∞ synchronization for complex dynamical networks with coupling delays using distributed impulsive control. Nonlinear Anal. Hybrid Syst. **17**, 111–127 (2015)
146. Y. Ma, N. Ma, Finite-time H_∞ synchronization for complex dynamical networks with mixed mode-dependent time delays. Neurocomputing **218**, 223–233 (2016)
147. P. DeLellis, M. DiBernardo, F. Garofalo, M. Porfiri, Evolution of complex networks via edge snapping. IEEE Trans. Circuits Syst. I: Reg. Pap. **57**(8), 2132–2143 (2010)
148. Q. Xu, S. Zhuang, Y. Zeng, J. Xiao, Decentralized adaptive strategies for synchronization of fractional-order complex networks. IEEE/CAA J. Automat. Sin. **4**(3), 543–550 (2017)
149. Z.G. Wu, Y. Xu, R. Lu, Y. Wu, T. Huang, Event-triggered control for consensus of multiagent systems with fixed/switching topologies. IEEE Trans. Syst. Man Cybern. Syst. (2017). https://doi.org/10.1109/TSMC.2017.2744671
150. X.F. Wang, G.R. Chen, Synchronization in small-world dynamical networks. Int. J. Bifurcation Chaos **12**(01), 187–192 (2002)
151. J. Lü, X. Yu, G. Chen, D. Cheng, Characterizing the synchronizability of small-world dynamical networks. IEEE Trans. Circuits Syst. I: Reg. Pap. **51**(4), 787–796 (2004)
152. Q. Zhang, J. Lu, J. Lü, C.K. Tse, Adaptive feedback synchronization of a general complex dynamical network with delayed nodes. IEEE Trans. Circuits Syst. II: Express Briefs **55**(2), 183–187 (2008)
153. L. Wang, H. Dai, H. Dong, Y. Shen, Y. Sun, Adaptive synchronization of weighted complex dynamical networks with coupling time-varying delays. Phys. Lett. A **372**(20), 3632–3639 (2008)
154. W.W. Lin, Y.Q. Wang, Chaotic synchronization in coupled map lattices with periodic boundary conditions. SIAM J. Appl. Dyn. Syst. **1**(2), 175–189 (2002)
155. P. Lu, Y. Yang, L. Huang, Synchronization of linearly coupled networks of deterministic ratchets. Phys. Lett. A **372**(22), 3978–3985 (2008)
156. S. Wen, S. Chen, W. Guo, Adaptive global synchronization of a general complex dynamical network with non-delayed and delayed coupling. Phys. Lett. A **372**(42), 6340–6346 (2008)
157. Y. Xu, W. Zhou, J. Fang, W. Sun, Adaptive synchronization of the complex dynamical network with non-derivative and derivative coupling. Phys. Lett. A **374**(15–16), 1673–1677 (2010)

158. D.H. Ji, S.C. Jeong, J.H. Park, S.M. Lee, S.C. Won, Adaptive lag synchronization for uncertain complex dynamical network with delayed coupling. Appl. Math. Comput. **218**(9), 4872–4880 (2012)
159. X. Guo, J. Li, A new synchronization algorithm for delayed complex dynamical networks via adaptive control approach. Commun. Nonlinear Sci. Numer. Simul. **17**(11), 4395–4403 (2012)
160. Y. Chai, L. Chen, R. Wu, J. Sun, Adaptive pinning synchronization in fractional-order complex dynamical networks. Physica A **391**(22), 5746–5758 (2012)
161. S. Liu, X. Li, W. Jiang, Y.Z. Fan, Adaptive synchronization in complex dynamical networks with coupling delays for general graphs. Appl. Math. Comput. **219**(1), 83–87 (2012)
162. X.Z. Jin, G.H. Yang, Adaptive synchronization of a class of uncertain complex networks against network deterioration. IEEE Trans. Circuits Syst. I: Reg. Pap. **58**(6), 1396–1409 (2011)
163. X.Z. Jin, G.H. Yang, W.W. Che, Adaptive pinning control of deteriorated nonlinear coupling networks with circuit realization. IEEE Trans. Neural Netw. Learn. Syst. **23**(9), 1345–1355 (2012)
164. X. Li, G. Chen, Synchronization and desynchronization of complex dynamical networks: an engineering viewpoint. IEEE Trans. Circuits Syst. I: Fundam. Theory Appl. **50**(11), 1381–1390 (2003)
165. X.F. Wang, G. Chen, Synchronization in scale-free dynamical networks: robustness and fragility. IEEE Trans. Circuits Syst. I: Fundam. Theory Appl. **49**(1), 54–62 (2002)
166. C. Posadas-Castillo, C. Cruz-Hernández, R.M. López-Gutiérrez, Synchronization of chaotic neural networks with delay in irregular networks. Appl. Math. Comput. **205**(1), 487–496 (2008)
167. X. Xu, Z Chen, G. Si, X. Hu, P. Luo, A novel definition of generalized synchronization on networks and a numerical simulation example. Comput. Math. Appl. **56**(11), 2789–2794 (2008)
168. Z. Fei, H. Gao, W.X. Zheng, New synchronization stability of complex networks with an interval time-varying coupling delay. IEEE Trans. Circuits Syst. II: Express Briefs **56**(6), 499–503 (2009)
169. Y. Fan, H. Liu, Y. Zhu, J. Mei, Fast synchronization of complex dynamical networks with time-varying delay via periodically intermittent control. Neurocomputing **205**, 182–194 (2016)
170. Y. Xu, W. Zhou, J. Fang, C. Xie, D. Tong, Finite-time synchronization of the complex dynamical network with non-derivative and derivative coupling. Neurocomputing **173**, 1356–1361 (2016)
171. X. Yang, D.W.C. Ho, J. Lu, Q. Song, Finite-time cluster synchronization of T-S fuzzy complex networks with discontinuous subsystems and random coupling delays. IEEE Trans. Fuzzy Syst. **23**(6), 2302–2316 (2015)
172. R. Xu, Y. Kao, C. Gao, Exponential synchronization of delayed Markovian jump complex networks with generally uncertain transition rates. Appl. Math. Comput. **271**, 682–693 (2015)
173. L. Wang, M.Z.Q. Chen, Q.G. Wang, Bounded synchronization of a heterogeneous complex switched network. Automatica **56**, 19–24 (2015)
174. J. Lu, J. Zhong, C. Huang, J. Cao, On pinning controllability of boolean control networks. IEEE Trans. Autom. Control **61**(6), 1658–1663 (2016)
175. J. Lu, Z. Wang, J. Cao, D.W.C. Ho, J. Kurths, Pinning impulsive stabilization of nonlinear dynamical networks with time-varying delay. Int. J. Bifurcation Chaos **22**(7), 1250176 (2012)
176. Y.W. Wang, W. Yang, J.W. Xiao, Z.G. Zeng, Impulsive multisynchronization of coupled multistable neural networks with time-varying delay. IEEE Trans. Neural Netw. Learn. Syst. **28**(7), 1560–1571 (2017)
177. X. Liu, Synchronization of linearly coupled neural networks with reaction–diffusion terms and unbounded time delays. Neurocomputing **73**(13–15), 2681–2688 (2010)
178. J.L. Wang, H.N. Wu, T. Huang, Passivity-based synchronization of a class of complex dynamical networks with time-varying delay. Automatica **56**, 105–112 (2015)

179. K. Wang, Z. Teng, H. Jiang, Adaptive synchronization in an array of linearly coupled neural networks with reaction-diffusion terms and time delays. Commun. Nonlinear Sci. Numer. Simul. **17**(10), 3866–3875 (2012)
180. S.X. Wang, Y.L. Huang, B.B. Xu, Pinning synchronization of spatial diffusion coupled reaction-diffusion neural networks with and without multiple time-varying delays. Neurocomputing **227**, 92–100 (2017)
181. K. Wu, B.S. Chen, Synchronization of partial differential systems via diffusion coupling. IEEE Trans. Circuits Syst. I: Reg. Pap. **59**(11), 2655–2668 (2012)
182. L. Zhu, L. Tian, D. Shi, Explosive transitions to synchronization in weighted static scale-free networks. Eur. Phys. J. B **86**(11), 451 (2013)
183. Q. Wang, Z. Duan, G. Chen, Z. Feng, Synchronization in a class of weighted complex networks with coupling delays. Physica A **387**(22), 5616–5622 (2008)
184. M. Porfiri, D.J. Stilwell, E.M. Bollt, Synchronization in random weighted directed networks. IEEE Trans. Circuits Syst. I: Reg. Pap. **55**(10), 3170–3177 (2008)
185. X. Wu, Synchronization-based topology identification of weighted general complex dynamical networks with time-varying coupling delay. Physica A **387**(4), 997–1008 (2008)
186. H.N. Feng, J.M. Li, Distributed adaptive synchronization of complex dynamical network with unknown time-varying weights. Int. J. Autom. Comput. **12**(3), 323–329 (2015)
187. L. Lü, C. Li, Generalized synchronization of spatiotemporal chaos in a weighted complex network. Nonlinear Dyn. **63**, 699–710 (2011)
188. G. Wen, W. Yu, G. Hu, J. Cao, X. Yu, Pinning synchronization of directed networks with switching topologies: a multiple Lyapunov functions approach. IEEE Trans. Neural Netw. Learn. Syst. **26**(12), 3239–3250 (2015)
189. W. Guo, S. Chen, A. Francis, Pinning synchronization of complex networks with delayed nodes. Int. J. Adapt. Control Signal Process. **29**(5), 603–613 (2015)
190. X. Liu, T. Chen, Cluster synchronization in directed networks via intermittent pinning control. IEEE Trans. Neural Netw. **22**(7), 1009–1020 (2011)
191. X.J. Li, G.H. Yang, Fuzzy approximation-based global pinning synchronization control of uncertain complex dynamical networks. IEEE Trans. Cybern. **47**(4), 873–883 (2017)
192. Q. Ma, J. Lu, Cluster synchronization for directed complex dynamical networks via pinning control. Neurocomputing **101**, 354–360 (2013)
193. C. Lin, Q. Wu, L. Xiang, J. Zhou, Pinning impulsive synchronization in coupled delayed directed dynamical networks, in *IEEE International Symposium on Circuits and Systems*, Melbourne, 2014, pp. 498–501
194. X. Li, X. Wang, G. Chen, Pinning a complex dynamical network to its equilibrium. IEEE Tran. Circuits Syst. I: Reg. Pap. **51**(10), 2074–2087 (2004)
195. T. Chen, X. Liu, W. Lu, Pinning complex networks by a single controller. IEEE Trans. Circuits Syst. I: Reg. Pap. **54**(6), 1317–1326 (2007)
196. G.S. Wang, J.W. Xiao, Y.W. Wang, J.W. Yi, Adaptive pinning cluster synchronization of fractional-order complex dynamical networks. Appl. Math. Comput. **231**, 347–356 (2014)
197. L. Guo, H. Pan, X. Nian, Adaptive pinning control of cluster synchronization in complex networks with lurie-type nonlinear dynamics. Neurocomputing **182**, 294–303 (2016)
198. M.A.A. Ahmed, Y. Liu, W. Zhang, F.E. Alsaadi, Exponential synchronization via pinning adaptive control for complex networks of networks with time delays. Neurocomputing **225**, 198–204 (2017)
199. R. Yu, H. Zhang, Z. Wang, J. Wang, Synchronization of complex dynamical networks via pinning scheme design under hybrid topologies. Neurocomputing **214**, 210–217 (2016)
200. C. Huang, D.W.C. Ho, J. Lu, J. Kurths, Pinning synchronization in T-S fuzzy complex networks with partial and discrete-time couplings. IEEE Trans. Fuzzy Syst. **23**(4), 1274–1285 (2015)
201. X. Liu, T. Chen, Synchronization of complex networks via aperiodically intermittent pinning control. IEEE Trans. Autom. Control **60**(12), 3316–3321 (2015)
202. W. Xia, J. Cao, Pinning synchronization of delayed dynamical networks via periodically intermittent control. Chaos **19**(1), 013120 (2009)

203. L. Xiang, J.J.H. Zhu, On pinning synchronization of general coupled networks. Nonlinear Dyn. **64**(4), 339–348 (2011)
204. Z. Xu, P. Shi, H. Su, Z.G. Wu, T. Huang, Global H_∞ pinning synchronization of complex networks with sampled-data communications. IEEE Trans. Neural Netw. Learn. Syst. **29**(5), 1467–1476 (2018)
205. Y. Guo, Exponential output synchronization of complex dynamical networks with time-varying delay via pinning control, in *Proceedings of the Fifth Instrumentation and Measurement, Computer, Communication and Control*, Qinhuangdao, 2015, pp. 10–14
206. C.X. Fan, G.P. Jiang, Adaptive pinning synchronization for complex network coupled with the outputs. J. Appl. Sci. **28**(2), 203–208 (2010)

FSC
www.fsc.org
MIX
Papier aus verantwortungsvollen Quellen
Paper from responsible sources
FSC® C105338

If you have any concerns about our products,
you can contact us on
ProductSafety@springernature.com

In case Publisher is established outside the EU,
the EU authorized representative is:
Springer Nature Customer Service Center GmbH
Europaplatz 3, 69115 Heidelberg, Germany

Printed by Libri Plureos GmbH
in Hamburg, Germany